EARTH WATCH

Notes on a Restless Planet

EARTH WATCH

Notes on a Restless Planet

by Jean and Daniel Shepard

1973

DOUBLEDAY & COMPANY, INC.
GARDEN CITY, NEW YORK

To
Anna C. Heck
Chester R. Heck

ISBN: 0-385-08325-4
Library of Congress Catalog Card Number 72–92241

Printed in the United States of America
First Edition

ACKNOWLEDGMENTS

Acknowledgment is made with deep gratitude to Robert Citron, director of the Center for Short-Lived Phenomena, Smithsonian Institution, for the confidence he displayed in our concept of a book he had envisioned, and for the extraordinary encouragement and cooperation extended the authors.

Our thanks also to the members of his staff in Cambridge, Massachusetts, David R. Squires, John Whitman, and Lee Cavanaugh for their interest and assistance; and special appreciation to Harold Banks, Jr., geologist, of the Smithsonian's National Museum of Natural History, Washington, D.C., for his invaluable efforts in identifying and helping to assemble illustrative material.

Jean and Daniel Shepard

Contents

Part V TABULATION OF EVENTS, 1968–73

List of Illustrations

Introduction

The world is still undergoing change under the influence of natural forces as old as time, and of man-made forces in competition with them. We orbit through space on an imperfect sphere whose thin skin is bombarded with celestial debris, ruptured from within by the pressures of a molten core, shaken by internal faults and stresses, swept by powerful tides, and implanted with countless life forms that exist in infinitely complex relationships to their environments and one another.

These basic forces of nature generate sudden, significant occurrences everywhere, ranging in impact from the curious to the disastrous. Among them are volcanic eruptions, earthquakes, the birth of new islands, the fall of meteorites and fireballs, and abrupt changes in biological and ecological systems. In their diversity, these natural events share two main characteristics: they are short-lived, and they are highly valuable as objects of scientific study before they have subsided or their effects have dissipated.

It was for this reason that the Smithsonian Institution formally established its Center for Short-Lived Phenomena in September 1968. The Center is located in Cambridge, Massachusetts, where it receives operational support from the nearby Smithsonian Astrophysical Observatory. It describes itself as ". . . an early alert system and clearing house for the reception and dissemination of information on short-lived natural events."

The Center collects and disseminates information as it is reported through advanced communications facilities by more than 2,700 registered correspondents, including scientists in many disciplines, organized into a 143-country network. These linkages on a round-the-clock basis are effected with such speed that research teams can often be mobilized to reach the scene with instruments and equipment while environmental changes are still in process.

Between 1967 and 1971, the Center documented 427 short-lived events, more than 250 of which were investigated by scientific field teams. The phenomena are divided into four categories of unpredictable natural occurrences which provide new field research opportunities for scientists:

1. Earth Science Events
2. Biological Science Events
3. Astrophysical Events
4. Urgent Anthropological[1] Events

Earth Watch follows the lead of the Center for Short-Lived Phenomena by presenting selected events under these headings. But this is not the way that they burst in on our awareness. Often unheralded, sometimes unexplained, they generate responses ranging from mild interest to justified panic. The chronicle of a single year should show how fortunate it was for primitive man's peace of mind (such as it was) that, while he was contemplating the natural mystery which had just staged itself on his own midden heap, he had no means of knowing what was happening simultaneously just over the mountain.

Within a single week in December 1967 undersea volcanic eruptions gave birth to two miniature islands, one in Antarctica and the second in the South Pacific. The first survived; the latter disappeared. At the turn of 1968 an earthquake in Sicily took 250 lives. In February a bright green fireball exploded above Polo, Illinois, while in March another passed over Dayton, Ohio, and three huge meteorites turned the night sky a brilliant white before crashing into a hillside near Veracruz, Mexico. In mid-April a smaller meteorite sailed over half the state of New York before landing with a noise like an exploding firecracker on the roof of a house in Schenectady.

Later in the same month Mount Mayon in the Philippines erupted after years of semiretirement; in May a violent earthquake in northern Japan wrecked shipping along the coast and destroyed homes inland. Shortly afterward, another earthquake convulsed the New Zealand landscape.

Mount Fernandino in the Galápagos Islands inaugurated the second half of 1968 with a stupendous air blast in June that shot a huge mushroom-shaped cloud so high it created its own electrical storm.

[1] The term given to discoveries from another age which must be studied quickly, before exposure erodes or changes them in some way. Examples of urgent anthropological events are finds of mastodon remains and the petrified footprints of primitive man himself. Encounters with new tribes living in prehistory also come under this classification, since their culture must be investigated before they are contaminated or destroyed by contact with the outside world.

Sicily's Mount Etna followed suit with crater explosions every three seconds, punctuating fiery streams of lava flowing down the cone. Overlapping this prolonged eruption, an earthquake in Moyabamba, Peru, killed 11 people and injured hundreds more, while higher up in the hemisphere, but much lower down on the disaster scale, a 55-mile-per-hour windstorm ravaged the foliage of hardwood trees in Vermont.

June and July were bad months for sea life and seaside dwellers. Mussels and sand eels off Newcastle, England, were fatally poisoned en masse by uncongenial organisms that bloomed spontaneously in the plankton on which they fed; an icy underwater tide sucked up from the depths of the Atlantic by Hurricane Brenda killed thousands of tropical fish and strewed their brightly colored bodies through 70 miles of coastal waters; the deaths of thousands of fish in northern California's Stanislaus River were attributed to unknown causes (although herbicides and pesticides were strongly suspected); an inexplicable sea surge struck the Valparaiso coast, causing substantial damage, and a seiche (or seismic disturbance of the waters) ruffled the surface of Lake Michigan.

There were more fireballs during the summer, over Omaha, Grand Rapids, Albany, and southern Argentina (however, the last was later identified as a satellite re-entry), a locust swarm in the Sudan, a drought in Chile, a flyaway of 10 million ducks down the Mississippi, and a mass migration of some 20 million Appalachian squirrels to all points of the compass in search of food.

Interspersed with these milder occurrences, Mount Kilauea on Hawaii erupted with fountains of lava, there were two earthquakes in Indonesia, one harmless submarine eruption in the Molucca Straits and a second in the Celebes Sea which killed 200 island residents. These were overshadowed by the violent, successive eruptions of Mount Arenal in Costa Rica, which began in late July and continued until the end of the year, initially wreaking havoc, death, and destruction over several square miles on its western slopes.

The first day of August brought disaster to Manila with a major earthquake that toppled buildings in the downtown business district and injured several hundred people. The next day a strong quake shook the sea bottom 175 miles east of Oaxaca, Mexico, causing no deaths, followed by one of the same magnitude in northeast Iran that caused 6,000 deaths and destroyed 60,000 homes.

In September an undersea volcano known as Banua Wuhu flurried the Celebes Sea and another, this one unnamed, set off a sand avalanche that cut a submarine cable east of New Guinea.

Maple trees along Massachusetts roadways began dying in general protest against the encroachment of man and his impurities, and on the last day of summer an earthquake in eastern Venezuela did considerable damage to the natural environment, leaving extensive fissuring in its wake.

Fall began with a substrata collapse in the Brazilian resort town of Guaratuba, owing to the infiltration of sea water; 20 buildings, including the city hall, were destroyed. New Zealand's Kermadec Islands had their second quake of the year; one in Turkey killed 22 people and injured 250, demolishing or damaging 2,300 homes; another severely damaged a number of buildings and caused numerous injuries in western Australia. The Philippines' Taal Volcano became active with a combination of increased surface temperatures and earth tremors, and Hawaii's Kilauea was heard from again in the form of eruptions accompanied by quaking.

A group of fur seals from the Bering Sea colonized San Miguel Island off Santa Barbara, California, in spite of the human presence, while Hawaiian monk seals vanished completely from the Midway and Kure atolls because of the human presence. A "red tide" killed fish off the Louisiana coast. New Guinea was shaken by two series of earthquakes, one in September and the second in October; at the end of the latter month a separate and distinct quake of even greater intensity occurred. A set of footprints left by a prehistoric man, and preserved by a fall of volcanic ash which then hardened, was uncovered by erosion in the vicinity of Chinandega, Nicaragua, whereupon the Cerro Negro Volcano in the same country went into a 43-day paroxysm. A fireball over Pocahontas, Iowa, saluted the month with a sonic boom and burst of light, permitted its trajectory to be filmed by five tracking cameras, and then dived to earth without leaving a trace at its calculated inpact point.

November began with a strong earthquake in southern Illinois and a falling meteorite in Portugal which struck the ground less than a hundred feet from a bystander. Mount Trident, an Alaskan volcano, gently cleared its throat and settled a coating of ash over the neighborhood for a distance of 60 miles.

December's contributions consisted of a sharp earthquake, with more than a hundred aftershocks, in southwestern Iceland; the last fireball of the year, which traced a bright path over Arizona; and the latest of a "swarm" of earthquakes centered offshore which had been jiggling dishes and windowpanes in the Santa Barbara area since the previous June.

Paralleling these natural events in the same twelve-month period were spills of oil and other toxic materials that tainted the ocean, coastal and inland waters. Off Puerto Rico and South Africa, in Virginia's St. James River, Long Island's Accabonac Harbor, Pennsylvania's Allegheny River, Delaware Bay, and the Gulf of Mexico, creatures of the sea, land, and air died in droves, smothered or immobilized in sticky oil or poisoned by pesticides and industrial wastes. Even the South Sargasso Sea turned up lumps of tar and gobs of petroleum during routine dredging by a research vessel to collect surface marine organisms.

These events are reported by the Center for Short-Lived Phenomena because of their catastrophic effect on the balance of nature. A single tanker breaking up in a storm can dump 5 million and more gallons of crude oil into the sea, creating a slick 60 miles square which then advances on the nearest beaches. Fighting concentrated pollution of this kind with detergents, fly ash, and other agents takes heroic efforts which often fail. The ecology of a sea-land area subjected to an oil spill can be upset for months, if not longer, as the various interdependent forms of life are driven away or killed outright. Further, oil can emulsify with water and sink to the bottom, blanketing the sea bed and eliminating it indefinitely as a factor in the local food-life generating cycle.

Earth seems to be able to heal itself of even the most grievous self-inflicted wounds from fire, flood, earthquake, and meteorological disturbances. Whether it can cope with the continual assault by man which is blighting its salt and sweet waters, land surfaces and air envelope is questionable. The oceans were once thought to be impregnable against the puny human waste products that lost themselves in their depths, the atmosphere was seen as a limitless vault into which lighter-than-air exudants would escape forever, the streams and lakes were perpetually self-cleansing, and the expanse of living space was inexhaustible. The fact that men could layer their planet with such incredibly complex material and technological superstructures as to endanger the underlying life support systems was unthinkable. Yet there are signs which point to the contrary, and oil spills are not the only ones, as two examples reported in 1972 show.

The first is known as the Oyster Creek Fish Kill. It occurred in late

January with the appearance of uncounted thousands of dead fish in a tributary of Barnegat Bay, a 30-mile-long arm of the Atlantic off the New Jersey coast. Ninety-nine per cent were menhaden, a virtually inedible relative of the herring and shad which is found from Nova Scotia to Brazil and farmed out of the sea as a source of oil, ground poultry feed, and fertilizer.

Although there were a few anchovies, striped bass, and bluefish among the victims, the federal Environmental Protection Agency (EPA) directed its investigation toward the reasons for the mortality of the menhaden, especially since live and apparently healthy shrimp, gobies, winter flounder, and blue crab were found in the area.

Adult menhaden (these were juveniles, three to six inches in length) are usually found off the Carolinas in January, since they prefer a water temperature of 59–68 degrees. It was theorized that the young menhaden had failed to migrate south because they had become acclimated to the artificially warmed water created by the thermal discharge of a new nuclear power plant, the Oyster Creek Generating Station. The poor fish had been swimming about happily in their fool's paradise when a sudden natural temperature drop killed them.

The generating station draws 460,000 gallons of water per minute from Barnegat Bay, uses it to cool steam, and then returns it to Oyster Creek fifteen to twenty degrees warmer. During the last two weeks in January 1972 drastic seasonal decreases in Oyster Creek's water temperature, wholly unrelated to the power plant, had been routinely recorded. One of these, on January 26, had been a sharp twenty-two-degree drop which was believed to be the undoing of the menhaden.

The second example is known as the Chino Farm Goose Die-Off. It occurred on the weekend of March 11, 1972, when some 2,000 of the geese being raised on the 6,000-acre Chino Farm, near Chestertown, Maryland, were discovered to be either dead or dying. Representatives of the U. S. Fish and Wildlife Service collected 150 dead and 50 sick geese for laboratory examination. From the symptoms of the sick birds it was thought that lead poisoning might be responsible, and 95 per cent of the autopsies pointed toward this cause.

The puzzle of how enough lead could have accumulated to harm so many among the goose population was solved by an investigation of the farmland. A 50-acre pond had been drained after a dam broke the previous fall. Years before, the pond had been heavily shot over by hunters, and a great concentration of lead pellets had rained down and sunk to the bottom. Scratching for sand and grit in the newly exposed lake bed, the geese had taken up bird shot into their craws and succumbed from a probable overdose.

In neither the Oyster Creek nor the Chino Farm events was man conclusively determined to be at fault. The finding was "not proven" in both cases, but the evidence points to an unwitting interference in the natural order of things with consequences that were disquieting, though relatively minor. A barely perceptible drop in the menhaden catch or a slight scarcity of roast goose around Christmastime is not likely to arouse apprehension. Yet it does reflect the fact that certain of the world's bounty is measured out in a delicate balance which can be easily upset by gross, man-made outpourings, whether of warm water or hot buckshot.

This is not to say that all ecological mistakes are made ingenuously. In debating the advisability of building a pipeline across the Alaskan tundra, its advocates stated that man was after all a part of nature, and that his needs must override the conservationist point of view. The United States Department of the Interior gave its approval to the pipeline, despite fears that its heat would melt the permafrost and its framework constitute a barrier to animal migrations. A great oil conduit snaking across one of America's last unspoiled spaces seemed a certainty until, in early 1973, the United States Court of Appeals blocked its construction. The decision was based on the narrowest of legal grounds, having nothing to do with the upsetting of nature's balance. A fifty-three-year-old law limited the right of way for pipelines crossing federal lands to a width of fifty feet. The oil consortium had plans for a right of way roughly four times as wide, not counting support facilities and service roads.

Although the Alaska case did not come to grips with the environmental issue, there were mutterings that do-gooders who put the birds and beasts above human needs had prevailed. In one New England state, citizen opposition to proposed nuclear generating plants on the seacoast drew an editorial attack from the leading-circulation newspaper. Environmentalists were accused of undue sympathy for the hermit crab and, in a more sinister vein, involvement in a subversive plot to prepare the country for a Communist take-over by frustrating solutions to the energy crisis.

However, it is not improbable that in Soviet Russia there have been editorials pointing the finger at officials suspected of subordinating the interests of the people to antelopes. The *Saiga tartarica,* a sheeplike member of an antelope subfamily, was once found in vast numbers throughout Russia, as far west as Poland. By the twentieth century it had dwindled into a single herd found in the Kazakh Soviet Socialist Republic, in Central Asia.

Soviet game laws were invoked to save the rare species from extinction and proved so effective that the saiga doubled to an estimated 1 million head between 1965 and 1971.

Having multiplied, the saiga next discovered the newly developed wheat lands of northwest Kazakhstan during their annual spring migration, arriving just in time each year to feast on the growing grain. They also brought with them the danger of spreading disease to the domestic livestock in the virgin lands.

Attempts were made to chase off the herds with helicopters, but they returned as soon as the craft disappeared. Aside from relaxing the hunting restrictions, the only course seems to be the installation of miles of expensive fencing along the migration routes. This was recommended in 1966 but never carried out because of the cost. Now the bedeviled authorities are confronted with the task of corralling at least 800,000 rampaging antelope into permanent pasture lands.

In Russia and the United States the future price of bread and electricity may well rise to cover the cost of maintaining large game preserves and safeguarding lowly sea life. Men who once saw themselves as completely at the mercy of natural forces which they could neither comprehend nor control no longer have that excuse. Much of the restlessness of the planet stems from the loss of innocence with respect to our part in altering certain natural processes, and from the conflict between the awareness that we can check the abuse of our environment and the unwillingness to do so.

We are alienating ourselves from our natural surroundings by goals which are being determined by rigid economic systems, by the lure of a better life through the unrestrained application of mass industrial technologies, and by decisions made for immediate political gain rather than long-range social benefit.

To the argument that man is a part of nature must be added the undeniable fact that he is the reasoning part, and therefore accountable. If his demands on earth's resources are so great and his contribution to their replenishment so negligible that he is a liability to both himself and his habitat, there is a remedy. It is the same one which his planet administered in ages past to the great reptiles and other life forms that lacked the ability to sustain themselves in the environment into which they had been born.

Part I
EARTH SCIENCE EVENTS

How to Tell One Volcano from Another While Maintaining a Respectful Distance

Erupting volcanoes are always spectacular, often beautiful, and sometimes approachable up to the very edge of the crater if conditions are right. They have been systematically observed long enough for a considerable body of scientific knowledge to have accumulated, together with its own specialized vocabulary. Much of this is Latinate in origin (lava, magma, bocca, lapilli), as is the name of the phenomenon itself, borrowed from the ancient crater which is still active on Vulcano, one of the Lipari Islands north of Sicily.

Volcanic eruptions differ in nature, depending in large part on the composition of the magma, or underlying molten rock material, which is believed to form at the relatively shallow depth of twenty or twenty-five miles below the earth's surface. Magma is charged with water vapor, gases, and other volatile substances, superheated and under pressure. The degree of viscosity of the basic rock melt, as well as the ratio of gases to solids, influences the kind and force of the eruption. It determines, among other things, whether the magma will be emitted as lava or as dust, or as a mixture of the many forms of volcanic ejecta that lie between these two extremes.

Hawaiian-type eruptions are the quietest, consisting mainly of a rise and flow of lava, at times in great volume. Gas content is low and bubbles out of the erupting magma easily, although the molten rock often shoots fountainlike into the air for hundreds of feet. The lava is very fluid (of low viscosity) and tends to pour out in thin, fast-moving flows that may spread great distances from their vents. Pyroclastic materials (fragments formed by fire) comprise only a small percentage of the total. An interesting though not ever present sidelight of the Hawaiian type of eruption is the welling up of a lava lake in the crater.

The Strombolian type of eruption is named for the famous and continually active volcano whose glow is always visible at night off the Sicilian coast. It is more explosive than the Hawaiian, owing apparently to the higher viscosity of the magma and greater gas pressure. Unable to escape easily into the atmosphere, the gas accumulates until it reaches an explosive point that throws magma into the air and shreds it. Where only moderately viscous material erupts, it is sufficiently plastic to take on a rounded or spindled shape in the air and fall back to the ground in the pyroclastic fragments known as bombs (angular shapes are called blocks). Simultaneously with these ejections from the crater, lava may flow from the foot of the cone, but not as the principal product as in the Hawaiian eruption.

The Vulcanian type of eruption is a step upward on the explosive scale. It may throw up a great many blocks which, because of the greater viscosity of the magma, were already solidified in the crater or not malleable enough to become rounded or spindled shapes in midair. The explosion is typically accompanied by a great cloud of ash-bearing gas. Lava flows are brief and thick, or lacking entirely.

The Vesuvian type of eruption occurs at the times when pressure is suddenly relieved on a column of magma while it is still deep in the volcanic conduit. This triggers a tremendous expansion (or vesiculation) of the magma, and the ash-laden, liberated gas escapes in great volume, and with great velocity, through the conduit, wearing away and abrading the walls as it does so. In the 1906 eruption of Vesuvius a huge column of gas rushed out almost vertically for twenty hours, reaching 6 miles into the sky. It also bored a cylindrical throat 1,500 feet across the cone. Quantities of fragmented material were also ejected, and lava poured out of the lower slopes. But the weight of the gas given off was estimated to exceed the liquids and solids by far, and illustrated the fact that there are instances when the abundance of gas is more important than the composition and viscosity of the lavas in determining the character of the eruption.

Plinian eruptions are of the variety, characterized by a stupendous outrush of gas, that Vesuvius underwent in A.D. 79, when Pompeii, Herculaneum, and Stabiae were destroyed. The name comes from Pliny the Elder, the natural historian who was then also prefect of the Roman fleet stationed at Misenum, on the northern end of the Bay of Naples, just above Puteoli (see "A Highly Civilized Crater: The Pozzuoli Uplift"). Pliny brought the fleet south to attempt a rescue of the people of Herculaneum but, arriving too late, put in at Stabiae. There he died from the effects of the poisonous fumes that overwhelmed the city from the still erupting crater.

In its Plinian eruption of 1906, Vesuvius threw out enormous volumes of pumice and ash in such great paroxysms as to cause a collapse of the summit, forming a caldera or bowl-like cavity, much larger than the comparatively small circular hollow of the original crater.

Such eruptions are based on vesiculation which spreads so rapidly through great volumes of magma that the rock melt is torn to bits by the resulting violent explosion. These were probably the conditions that existed in the greatest known volcanic eruption of historical times, that of Tambora, on the Indonesian island of Sumbawa, in 1815. The explosions blew out 35 cubic miles of debris over the surrounding area, and the collapse of the cone reduced the height of the mountain from 13,000 to 9,000 feet. In its Plinian eruption of 1905, Vesuvius lost 1,600 feet of its 4,275-foot summit.

On the top rung of the ladder of destructiveness are the Pelean eruptions. Their principal features are the nuée ardente (glowing cloud) (see "A Crater Keeps Cool: Soufrière Volcanic Activity") or the closely related froth flow, both of which are similar in mechanism to the Plinian eruptions and sometimes associated with them.

The Pelean eruptions, named after Martinique's Mount Pelée, characteristically begin with the growth of a volcanic lava dome, accompanied by a superheated avalanche that descends at speeds as high as 100–160 miles per hour and raises great boiling clouds of dust as it comes. (Froth flows are equally mobile masses of incandescent pumice and pumiceous ash.)

Each of the particles in the avalanches and the clouds above them is actually in the process of liberating gas. Each expanding gas envelope pushes against its neighbor, creating an enormously growing, virtually frictionless anti-atmosphere that chars, suffocates, and buries everything in its path. This is what happened to St. Pierre and its 30,000 inhabitants in a matter of seconds during the 1902 eruption of Pelée.

Perversely, volcanoes do not adhere cleanly to these classifications. Overlapping types of eruptions are as much the rule as the exception. A single volcano may exhibit some or all of these kinds of activity in successive eruptions, or even during the same eruption from different vents. The only consistency is that, during a major eruption, there is not much to choose from in terms of peril by those who happen to be within range.

Among the three to four hundred active volcanoes throughout the world, here are a handful that reflect the range of personalities displayed by the highly individualistic members of this fraternity.

On Saturday, April 20, 1968, the 8,000-foot conic volcano, Mount Mayon, located 200 miles southeast of Manila on the island of Luzon, Republic of the Philippines, began spewing steam and molten lava and shooting balls of fire 3,000 feet high. The last eruption of Mayon had occurred in 1947.

By Monday, April 22, activity in the volcano zone had become so intense that evacuation was started of 70,000 residents within the 6-mile danger zone. Lava flow began in the early morning of the next day, characterized by strong fountaining during eruptions and quiet overflowing from the crater in between.

Eruption explosions intensified during the week of April 22, and by Wednesday, April 24, more than twenty separate eruptions had been recorded. By the morning of the same day, a pyroclastic cloud had risen from the cone, reaching 30,000 feet during the week of April 28.

On April 30, with the permission of the government of the Philippines and the assistance of the U. S. Air Force, scientists of the Smithsonian Institution of Petrology filmed the eruption from the air. The scientific team then spent nine days on the volcano, observing eruptions, collecting samples, and directing the film crew. Several spectacular nuées ardentes were witnessed and photographed by the ground team, while the Air Force flew missions during the next six weeks of the eruption.

On May 5, eruption activity subsided, with sporadic weak ash explosions and lava emissions continuing to the end of the month.

At 4:21 P.M., Greenwich Mean Time, on June 11, 1969, the seismic station at Quito, Ecuador, recorded shocks originating in the Galápagos Islands, the Pacific archipelago 650 miles to the west which belongs to Ecuador. At 5:00 P.M. local time on the same date, a tremendous explosion occurred on one of the Galápagos group, Fernandina Island. The shock was felt on Santa Cruz Island, 90 miles to the east, and was heard on San Cristóbal Island, 150 miles away. The air blast was described by infrasonic detection stations in North America as "stupendous and in the multi-megaton range." Seismic stations recorded continuous tremors from 4.5 to 5.5 on the Richter scale.

A mushroom-shaped cloud rose high in the sky to the west of Santa Cruz, and at 5:45 P.M. a number of further explosions occurred. The cloud, a spectacular sight in an otherwise clear sky, extended the 90 miles to Santa Cruz. There were flashes of electrical discharge over the area during the night, and a fallout of gray volcanic ash reached Isabela Island 56 miles to the southeast. Continuous seismic activity was recorded for the next few days.

On June 19 a party from the Darwin Research Station reached the rim of the Fernandina caldera. Their report states in part:

"During an ascent to the rim of the caldera, continuous tremors were felt. The frequency and violence of these tremors was such that, at one point on the main outer slope of the volcano, at an altitude of 685 meters, between the six hours 1730 to 2330 local time (5:30–11:30 P.M.), 56 were counted, each lasting from two to six seconds, and of these 14 were sufficient to cause rock falls from cones on nearby slopes. Trees were seen to be shuddering as if a strong wind were blowing through their branches. The tremors grew and subsided, rather than being felt as abrupt shocks, giving the impression almost as if the whole island were balanced on a mass of jellylike material. It was found that these movements had caused, and were continuing to cause, the collapse of the inner walls of the caldera; tremendous falls of rocks occurring at frequent intervals and giving rise to great clouds of dust that filled the crater and overflowed in the direction of the prevailing winds."

The sound of the falling rocks was likened to the roar of heavy seas breaking on a rocky beach. A strong wind arose after each fall. Large areas of the edge of the caldera had fallen into the crater, and there were heavy fissures in parts of the existing caldera border. Cliffs and trees had been dislodged in places by the tremors. Because of the heavy clouds of dust, it was not possible to see the floor of the crater. It had contained a lake which formed a complex of islands and inlets owing to the unevenness of the caldera bottom. Bordered with reeds, it had supported probably the largest population of indigenous ducks in the Galápagos.

Vegetation over the entire island was covered to a varying extent with volcanic ash. Because of the dramatic disturbances which might have occurred to various forms of life within range of the eruption, several dozen biologists around the world who were known to be interested in the ecology of the Galápagos Islands received notification of the event.

During the week of June 13 the Center for Short-Lived Phenomena heard from a number of seismic and infrasonic stations in South America and the United States that the air blast of June 11, as well as subsequent seismic activity, was of an unusually large magnitude. The Santa Cruz station reported continuous seismic activity between June 16 and 22, with an average of two hundred shocks per day, and the largest number on June 19.

On June 24 the Center requested U. S. Air Force transportation for a team of scientists. The request was granted, and on July 3 a group of seven scientists and observers flew to the Galápagos to study the effects of the eruption on the physical and biological environment.

On flying over Fernandina on July 4, the team reported that ". . . the southern part of the caldera floor has been lowered by approximately 300 meters, the old central cone on a now-tilted floor remains intact and the large lake was still present."

By July 8 seismic activity had all but ceased. Nothing of further interest occurred until almost three years later when, from April through June 1971, a swarm of earthquakes centered near Fernandina were recorded. The magnitude and pattern were similar to the caldera collapse of 1968, and it was believed that it had resumed. But an inspection of the island, and nearby islands with calderas, revealed that no such activity had taken place, and the great explosion of 1968 remained for the time being without sequel.

Mount Etna, on Sicily's east coast, is the highest volcano in Europe and the highest mountain in southern Italy (10,700 feet). Its elliptical base measures 31 by 19 miles and covers 460 square miles. A truncated cone in shape, it has a great rift three miles in width, known as the Valle del Bove, on its east side. This is the original crater and is no longer active. The present crater is at the summit, and there are more than 200 subsidiary cones, some over 3,000 ft. in height, which have risen on its slopes.

Etna is mentioned by Thucydides as having been in eruption as far back as the eighth and fifth centuries B.C. Most recently, it has been active from January 1966 on, characterized by explosions of varying intensity within the northeast and central summit craters and by slow emissions of lava on the upper slopes of the northeast cone. This persistent but generally moderate activity culminated in the massive outbreak of 1971.

The eruption began on April 5 of that year, with the opening at the foot of the central crater of two radial fractures about 200 meters (650 feet) apart, one directed toward the so-called eastern vent, elevation 3,100 meters, (about 10,200 feet) and the other toward the western vent, at 3,000 meters (about 9,900 feet). Continuous explosive activity quickly built new cinder cones 30 to 50 meters high on the fissures. Meanwhile, lava flows poured out of the lower part at initial speeds of up to 15 kilometers (about 9 miles) per hour.

During April, lavas flowed on the south flank of the mountain, known as Piano del Lago, destroying the observatory and the upper part of the cableway. There were also lava flows in the southeasterly direction through the Valle del Bove, and more advanced lava fronts toward the south and southwest reached the 2,200-meter level.

On the evening of May 4, a new fissure opened in the east base of the central cone, about 2,900 meters above sea level. The discharge of highly fluid lava increased during the night. The strong degassing, with the emission of lava lumps and incandescent scoria (loose fragments of frothy material), built an elongated, sharply peaked cone 30 to 35 meters high. The cone was open toward the east, where a large amount of lava was flowing with an initial speed of 10 to 15 kilometers per hour.

Lava flows began to decrease, and by the morning of May 7 all was quiet on the crater except for some small, cooling lava flows at the south vents and some increased emission of gas from the central crater, accompanied by rumblings and brown cinder clouds.

That night, however, a large red glow could be seen toward the east on the Valle del Bove. A new fracture originated from the upper east vent and reached down to the 2,400-meter level. There, a flank eruption started on the slope between the Valle de Leone and Valle del Bove (the 2,700-meter to 2,400-meter level). A half dozen lava flows were vigorously active, but explosive activity was almost negligible. This eruption continued the next day while continuous but relatively mild degassing was building small eruptive conelets (hornitos).

On the night of May 11 a group of people standing near Rifugio Citelli, on the east-northeast slope at an elevation of 1,750 meters, heard a tremendous noise. They thought at first that it originated in the upper regions of the mountain but a short time later observed a growing glow only one kilometer away. It was the eruptive fracture of a few days before, which had reached the 1,800-meter level and was now 5 kilometers long.

For three weeks, starting with the morning of May 12, huge quantities of lava were delivered from this new center of activity at a more or less constant rate. The erupting magma was highly fluid, with an initial speed of 20 to 25 kilometers per hour, but no new explosion occurred. Consequently, no new pyroclastic cones were created.

The lava flows invaded the pinelands of Cubania and spread to the rich lower eastern slopes of the mountain, seriously threatening the villages of Fornazzo and Sant'Alfio. They stopped in the valley of Cava Grande, however, at an altitude of only 600 meters, the total length of flow having reached 8 kilometers.

Early on the morning of May 18, an explosive outbreak started from a circular, newly opened abyss at the upper end of the fracture, near the 3,000-meter level. For many days high-pressure gases, throwing cinders and blocks, continued to escape from this crater, which was the actual explosive vent of this flank eruption.

During the first days of June the discharge of lava at the vents above Citelli began to decrease, and on the twelfth day of the month the eruption ended. Measurements of the lava showed the temperature to be extremely high (1,120–1,130 degrees centigrade), of extremely low viscosity, and with fewer crystals in its composition than had been found in previous eruptions. These factors explained its fluidity and rapid flow.

After a hundred days of complete quiescence, volcanic activity began again. The Istitutio Internazionale di Vulcanologia, of Catania, Sicily, reported on September 23 that a lava column raised up 200 meters above the rim of the main crater, which is known as La Voragine. Two explosive boccas (eruptive orifices) were seen at the bottom of the crater. This eruptive activity lasted only a day or two and ended suddenly.

Activity commenced once more in December 1971. From the two La Voragine boccas, volcanic sands and ashes mixed with fragments of molten lava were thrown 400 meters into the air, falling into the surrounding area within a radius of about 300 meters. And so Mount Etna ended the year.

Mount Arenal, in northwestern Costa Rica, began a series of violent and devastating Pelean-type eruptions on July 29, 1968, wreaking havoc, death, and destruction over several square kilometers on its western slopes and flank. A nuée ardente produced deadly blasts of hot gases, ash, and incandescent blocks of rock and lava. Some of the blocks were directed laterally and traveled swiftly down the mountain, killing or laying waste everything in their path. Explosive activity subsided by late August, but fumarolic activity continued into September. Late in that month a lava flow was reported, and on November 15 a molten stream 200 meters in length and 20 to 30 meters in height was observed traveling at an approximate speed of 1.5 to 2 meters per day.

Lava continued to flow from Arenal during 1969, and early in that year a 100-foot-thick flow of aa (lava heavily interlarded with solid materials) moved down the west flank of the volcano. It could be observed from a hill on which a monument had been erected to the 80 people who died in the previous year's eruption. By late 1969 the flow completely surrounded the memorial hill.

Activity throughout 1969 was usually accompanied by vapor eruptions, and on May 3 there were simultaneous outbreaks of Arenal and two other Costa Rican volcanoes, Poás and Rincón de la Vieja. For the remainder of the year Arenal evidenced massive lava flows toward the west. Initial speeds were slow, five feet per day being typical, but the flows had heads 60 to 100 feet high; one of them built a continuous mile-long, quarter-mile-wide surface of aa with curving, transverse flow ridges on its lower half.

The eruption continued into 1970 but was very quiet. Movement of flow ceased on August 29 and did not resume, although an even glow could be seen near the vent. The flow area, as measured on September 11, was about 3.2 kilometers, with an approximate volume of 48 million cubic meters. The mountain slumbered but continued to fume.

As volcanoes measure their age, Nicaragua's Cerro Negro was born yesterday. It created itself in 1850 and since then has earned the deserved reputation of Central America's most active volcano. A typical cone of ashes, it stands approximately 635 meters above sea level and 235 meters above the surrounding plain. These heights are subject to change during the numerous eruptions it has displayed over the years. These have been accompanied by lava overflows which enlarge its circumference and by the appearance of a large number of adventitious cones produced as the result of eruptive activity.

In a recent series of outbreaks, the volcano began to erupt on October 24, 1968. There were intermittent moderate explosions ejecting blocks, incandescent material, and a dense ash-smoke cloud to altitudes of several thousand meters.

A flank eruption of a double vent, a few hundred meters from the main cinder cone, was characterized by fountaining lava and an aa flow that began simultaneously with that of the principal volcano. This double vent quickly built itself into a small cinder cone and continued, together with Cerro Negro, in a state of continuous activity for a period of forty-three days.

During this period, October 24–December 7, Cerro Negro produced an eruption cloud which reached altitudes of 5,000 and 10,000 feet. The lava flow from the small vent reached a length of approximately 2,000 meters and a height of 10 to 20 meters by December 7. Occasional bursts of smoke and ash from the main vent, and a reduced volume of ash from the new vent, continued through the month of December.

The young volcano held its peace for almost exactly one year. Then, over a ten-day period from about December 19 to 29, 1969, it awoke with a series of mild gas and ash explosions. Each lasted only about one minute but blew a small ash cloud 2,000 feet high. As observed from the air, the crater had become somewhat enlarged since the previous year. The outbreak subsided from four or five explosions per day to two and then only one on December 29.

This was only the prelude to a major and prolonged eruption which began on February 3, 1971. A continuous, violent emission of gases, bombs, stone blocks, ash, and lapilli (fragments less than 2.52 inches in diameter) took place. A cloud reached to a height of 20,000 feet before being dispersed by the prevailing winds. During the first days of activity, the explosions came every ten seconds or so and formed continual turbulent clouds. Between the cloud and the main cone, continuous electrical phenomena could be seen, possibly caused by static charges generated by the rubbing together of particles and by the great temperature differences existing within the cloud. The pyroclastic materials were thrown considerable distances from the crater (approximately 500 meters) and generally in an easterly direction.

On February 9 the eruption changed character, passing from a continual to an intermittent phase. The cycles in the early days were fifteen minutes of activity and ten minutes of calm; the periods of calm became longer as the days passed, until the total cessation of activity on February 14.

An area of 5,785 square kilometers was affected by falling ash, with 785 square kilometers receiving a fall of ash more than 2.52 millimeters in thickness. The total volume of the ash thrown was 45,160,000 cubic meters.

The 1971 eruption of Cerro Negro was one of the most violent in recent times, possibly because no openings of adventitious craters were produced in the base or flank of the main cone. For this reason, energy was retained completely within the principal crater, and the eruption came totally from this source. It was only during the sixth day that a small amount of activity was observed in an adventitious crater which dated from 1957 and is almost totally covered by pyroclastic materials. However, this was not sufficient to relieve the pressures within the main cone.

As a result of the eruption, there was a great widening of the main crater, from 150 meters at its greatest diameter to 400 meters. At the same time there was a widening of the base of the cone from 620 meters to 1,100 meters. There was also a general increase in height and the partial or total concealment of numerous old vents which had appeared in 1923, 1947, 1957, 1960, and 1968.

Cerro Negro is obviously growing, as befits a youthful volcano a mere 122 years old, though not as fast as Paricutín, to the north in Mexico. Within eight months after Paricutín made a sudden appearance in a cornfield in 1943, eruptions built the cone to more than 1,500 feet and covered an area of more than 10 square miles with lava. Such instances are few and, fortunately for cartographers, far between.

Pyramid from Under the Sea

(The Kovachi Submarine Eruption: South Pacific, 1969–70)

On October 28, 1969, a Solomon Islands Airways pilot reported that, while on a flight over the sea southeast of New Georgia Island, he had seen volcanic activity. The site was Kovachi, in the British Solomon Islands area, a submarine volcano whose last eruption was noted in 1961 and which was in the habit of forming small, disappearing islands.

The Geological Survey of the Solomon Islands placed the hide-and-seek volcano under surveillance and, between October 29 and December 24, observed a number of water eruptions. Typically, they lasted from thirty seconds to one minute, were from 20 to 300 feet in height, and twice were accompanied by steam clouds. Most were moderate in magnitude, with one classed as slight and another bordering on severe. In between eruptions, the water boiled and bubbled, and the surface of the sea for a radius of 70 miles was streaked with red-brown scum.

As of late December, the activity was still continuing. On the twenty-second and twenty-fourth of the month, waterspouts were rising to 100 feet every five to ten minutes.

Observation was interrupted by the Christmas season and did not resume until after the New Year. In the meantime, Kovachi devised its own nativity celebration.

On January 30, 1970, the cargo vessel *Fortune Glory* observed that a small island bearing a cone had appeared and was in full volcanic eruption. From five miles distant, black smoke was seen rising, with glowing rocks being hurled upward. The cone was red hot at its crest, and steam rose from the sea all around its base.

The eruption persisted through February 6 and ceased on February 7, when a close examination was made. A pyramid of black rock 20 feet high remained, still hot enough to convert the waves breaking over it into steam. The cone seemed massive, with the remnants of a vent 5 feet in diameter on its northeastern side. The inner walls of the vent were coated with ocher, yellow, and red-brown material.

A week later the cone was still in existence while a Geological Survey ship delayed by three cyclones in the area prepared to make a visit. Two days later, as the ship approached, it was to find that the cone had sunk beneath the surface.

Before subsiding, the island had a surface area of approximately 500 square feet and appeared to be breaking down because of severe wave action. No lava samples could be obtained at that time because of rough water. Subsequent to February 16 the cone could be clearly seen just below the water, and on March 3 some lava specimens were dredged off the submerged slopes.

On that visit it was seen that hot water and mud were rising from the central vent and being carried off by the prevailing current. The mineral salts brought up in this "plume" of material had obviously increased the plankton, and there were large schools of fish and several sharks in the vicinity.

Continual observation was kept on Kovachi for the remainder of the year. It continued to emit muddy water, especially during October when the rate of emission grew to extend the discoloration of the sea over a large area. Underwater explosions were also reported on several occasions.

When, in December 1970, the Geological Survey ship again visited the volcano, there were no underwater explosions and no significant change appeared to have taken place, except for one: the fish and shark population had increased considerably, both in numbers and in size. In its small way, Kovachi is parodying the creation of other land masses of volcanic origin which rose out of the sea and became the focus of the life forms that would one day make it habitable.

Beneath the waves, the island awaits the day when it may reappear, perhaps this time to stay.

A Notorious Volcano

(The Hekla Eruption: Iceland, 1970)

Nature and Hollywood have created so many volcanoes in the tropics that both fact and fiction tend to obscure an important characteristic concerning the great pressures deep within the earth: they have no particular preference as to which point of the outer crust they breach in order to seek an exit.

There are, for this reason, numerous active volcanoes in parts of the world where skis rather than catamarans are the most practical means of travel. Iceland, unlikely as it may seem, is built up almost entirely of volcanic rock and has more than a hundred active volcanoes. One of the most notorious is Hekla, a 5,100-foot peak which began an eruption on May 5, 1970, that lasted for exactly two months.

During the first phase of the eruption the cloud of tephra, or volcanic ash, reached 5,000 feet. The main tephra fall lasted two hours, averaging 10,000 cubic meters per second, and the total volume was about 70 million cubic meters.

Hekla's deadly reputation stems in part from the high proportion of fluorine with which its tephra clouds are polluted. This is the same gas which was present in the monumental eruption of Pelée, on North Martinique, in 1902, that killed 40,000 people in a suffocating, charring cloud.

Falling over an area of some 25,000 square miles, (Iceland is only 39,709 square miles in total area) the fluorine-bearing ash cloud from Hekla caused the deaths of some 1,500 ewes and 6,000 lambs.

The real Hekla fissure, which splits the ridge lengthwise, did not open on this occasion, except at its extreme southeast end. Several smaller fissures at lower levels did, however, and they were sufficient to produce 11.5 square miles of lava flow.

Iceland is fortunate in not having any of its volcanoes in constant activity. There are often long intervals of quiet, broken by solitary outbreaks. With a volcano like Hekla to poison the tableland above which it rises, this is the best that can be hoped for. In 1783 the eruption of another crater, Laki, caused the loss from disease and starvation of 11,500 cattle, 28,000 horses, and 190,500 sheep through the destruction of their pasturage by volcanic ash. This was well over 50 per cent of all the livestock on the island, and Iceland was then visited by a famine that destroyed 9,500 people, or one fifth of the total population.

The Awakening of Cheldag

(The "Mud Volcanoes" of Azerbaijan: U.S.S.R., 1970)

After a hundred-year sleep, a mud volcano named Cheldag awoke in June 1970. Located seventy kilometers west of the Caspian seaport of Baku, in the Russian Soviet Socialist Republic of Azerbaijan, Cheldag heralded its renaissance with a powerful emission of gas. Soviet scientists watched as a cupola-shaped hill composed of cementlike fragments of dry clay appeared in the center of the volcano's crater. The dome gradually expanded in size to a height of 5 or 6 meters and ten times as much in diameter.

A month later another mud volcano, this one underwater in the Caspian ninety kilometers south of Baku, formed an island 55 by 105 meters in area and 1.5 meters above the surface of the waves.

Baku is one of the richest oil-producing centers in the world and the two mud volcanoes were probably of the false variety (the true ones are allied to geysers) which have nothing to do with vulcanicity but are caused by gas escaping from petroleum-bearing strata. The Caspian, together with the Crimea and Burma, is well known for this kind of phenomenon, and four were reported there during 1970 alone.

A Soviet academician was of the opinion that the growing activity of Azerbaijan mud volcanoes in recent years might possibly be connected with mountain formation, which is increasing all over the globe in the geological regions known as the folded Alpine mountains.

A Crater Keeps Cool

(Soufrière Volcanic Activity: St. Vincent Island, West Indies, 1971)

The West Indian island of St. Vincent is all of eighteen miles long and eleven miles from coast to coast at its widest point. With a total area of 133 square miles, it hardly seems cut out to accommodate an outsize, and somewhat unpredictable, volcano. Yet at its northern end looms the 4,018-foot summit known as the Soufrière. The name, transliterated from the original Italian, means the same as solfatara (see "A Highly Civilized Crater"), or a crater which has given up violent eruption as a way of life and taken to continuous fuming instead. For St. Vincent's Soufrière (there are others in the Windward Islands), the designation is a charitable one. It was not known historically prior to 1498 when it was said to have been discovered by Columbus, but since then it has by no means been quiescent. There were violent outbreaks in 1718, 1821, and 1902–3, the last of the series beginning synonymously with that of Mount Pelée on Martinique, which lies to the north.

From 1946 on, the temperature of the water in Soufrière's crater lake was measured monthly. In that period it was found to be about four more degrees above centigrade than would normally be expected from the temperature of the surrounding air. It was postulated that the additional heat came from small steam vents in the lake bottom. Apart from this, and from the occurrence of a swarm of local earthquakes between November 1945 and February 1946, the Soufrière remained completely dormant—until October 1971.

September 19, 1971, was the last date on which the crater was visited by an experienced observer and found to be completely normal. An aerial photograph, taken nine days later on September 28, showed no steam or discoloration of the lake water.

The first witness to abnormal activity was a visitor to the island who climbed to the crater rim on October 17 and noted steam and a sulfurous smell rising from the lake surface. He did not recognize the significance of the situation, however, and his account did not reach qualified authorities until the beginning of November. He had also taken photographs, which were compared with pictures made earlier in 1971 and showed that the lake level had risen about 8 meters by October 17, while the amount of visible steam signified a water temperature of not less than 60 degrees centigrade.

Almost simultaneously, the general alarm was raised when the pilot of a light plane reported to government officials that the lake was steaming. This was verified by further inspections from the air, and scientists from the Seismic Research Unit, University of the West Indies in Trinidad, were called to St. Vincent.

During the 1902 episode the inhabitants of Martinique, alarmed at the preliminaries being staged by Pelée, were reassured on hearing that St. Vincent's Soufrière was also in eruption, believing that this would relieve the pressures on their own crater. Shortly thereafter, however, Pelée devastated one tenth of the island and its chief commercial center, St. Pierre, with a dread nuée ardente, a huge incandescent avalanche of glowing volcanic ash and dust that cremates everything in its path and smothers all life in the vicinity with lethal gases. Forty thousand people died.

Meanwhile, St. Vincent was experiencing its own troubles from the Soufrière. The 1902–3 eruption emitted an estimated 1.4 cubic kilometers of fragmentary material, including ash showers and nuées ardentes, laying waste the northern third of the island and killing 1,565 people. The majority of the deaths occurred on the eastern, and most densely populated, flank of the mountain. Before the residents of this area were fully aware that an eruption had begun, strong explosions blew out the crater lake, generating hot mudflows which filled the major river valleys and cut all routes leading southward. The climax and glowing avalanches followed four hours later.

After this eruption, a lake re-established itself in the summit crater, and an understandably apprehensive watch was kept on Soufrière. In the subsequent decades, both the level and the temperature of the water has been measured, since a rise in either would warn of a possible outbreak. Between 1910 and 1971, the level remained constant within a few meters and the temperature within a few degrees.

The scientific team made its first descent to the crater lake on November 3. The water level was found to be 34 meters above normal and the temperature at the lake edge 81.5 degrees centigrade. From the change in water level, it was calculated that the eruption had begun in the second week in October and that new lava approximating 30 million cubic meters had been emitted into the lake between then and November 3 at a rate of not less that 1.2 million cubic meters per day.

From then on, the lake was visited at intervals of at least two days. Between November 3 and 20, the water level continued to rise at a diminished rate, and the temperature remained almost constant, fluctuating from 79 to 82 degrees C. On November 20 the last doubts about the emission of lava were dissipated when lava islands emerged from the water surface at several points slightly south of the lake center and about 210 meters above its former floor. Up to that time there had been the possible alternative of large-scale fumarolic activity in the form of an enormous volume of steam rising through the lake floor and condensing to heat the water and add to its volume. The appearance of the islands, however, confirmed a true eruption involving fresh lava.

The lava islands continued to grow. The first were small islets within an elongated area of about 100 by 300 meters. Three prominent east-west ridges then rose to the southeast of the original area. By December 1, 1971, which was the first date that the crater was reasonably clear of cloud and hot vapor, a pattern of ten radial ridges, or lobes, had developed within an area 300 meters in diameter.

Between December 1 and 8, the water within this circle was progressively displaced by rising lava, while the diameter of the island area increased by 450 meters, mainly by lateral growth in a westward direction. From December 8 onward, the island expanded steadily upward and outward. By January 31, 1972, it had reached a diameter of 610 meters and a height above current water level of 71 meters.

Until the appearance of the new lava on November 20, the level of the lake water had risen about 15 centimeters per day. After this date, however, it began to drop at the same rate until the end of November, at which point the drop accelerated to 30 centimeters per day. The water loss was calculated at 50 million cubic meters, almost all by evaporation. The total mass of new lava was estimated at 80 million cubic meters.

Surprisingly, the temperature of the water, which fluctuated between 77 and 82 degrees centigrade through the month of November, began to fall, with only minor variances, until it had reached 59 degrees centigrade on January 31. The decrease, in spite of the lessened water volume and apparently constant rate of lava emission, was attributed to the lava itself. The hot core of the lava dome was gradually insulated by the increasing thickness of the cooled outer shell.

The seismic activity associated with the eruption was very slight. To a permanent seismographic station 19 kilometers south of the crater was added a second 6 kilometers southwest, and a third 6 kilometers east. They detected nothing of any real significance.

When on November 20, 1971, however, it was determined that fresh lava was rising into the crater, it was decided to establish a network of radio-linked seismic stations closer to the summit. Accordingly, three telemetric stations were installed on the flanks of the mountain, one only 250 meters from the southwest rim of the crater. The timed signals from these stations indicated that such tremors as there were originated at depths of no more than one kilometer.

The virtual absence of earthquakes originating at greater depths was interpreted to mean that no really large volume of molten rock was on its way to the surface, and this in turn meant that no major eruption was expected.

The 1971–72 eruption, was a mild one involving the growth of a volcanic dome. It was completely different in character from the three earlier historic eruptions of the Soufrière, all of which were violently explosive throughout. It was, however, similar to the second half of the two large historic eruptions of Mount Pelée, Martinique, in 1902–5 and again in 1929–32. In each of these outbursts there was an initial, highly explosive phase which lasted several months. This was followed in each case by a second phase in which fresh lava welled up into the crater to build a dome. In both Martinique eruptions, dome growth continued for more than two years. Comparable periods of between six months and three years have been reported for dome building by volcanoes in other parts of the world.

Estimates were that the latest St. Vincent eruption would continue for a year, or possibly even two. This conclusion was based on the constancy of the rate of lava emission, about 500,000 cubic meters per day, and by the comparatively small total volume erupted to that point (80 million cubic meters as against 2.6 billion cubic meters during 1902–3).

The likelihood that the 1971 eruption will ever develop into dangerously explosive activity is remote. In at least 95 per cent of the historic eruptions at similar volcanoes, explosive activity began within the first two months of the onset of eruption. At last reports, Soufrière's nine-month eruption brought it comfortably past the initial period of high uncertainty.

At the estimated rate of lava emission, the summit of the new lava dome should reach the level of the lowest parts of the crater rim by May 1972, and the crater will be almost completely filled by 1973. The consequences of the filling of the crater were not seen as important. The viscosity of the new lava is so high that, even if the dome were to grow so that its flanks overlap the present crater rim, it is unlikely that the molten lava would escape down the mountain.

"We Are Having a Jökulhlaup from Grímsvötn"

(Glacial Outburst Floods: Iceland, 1972)

Iceland has more than 120 glaciers, and some of them sit atop its more than 100 active volcanoes. This gives rise to an unusual anomaly composed of fire and ice: the jökulhlaup, or glacial outburst flood. When an eruption occurs under one of these icecaps, the snow and ice melt and sometimes give rise to serious and sudden floods. In 1362 waters from the highest of the country's mountains, 6,424-foot Oerafajökull, in this way bodily swept 40 farms together with all their inhabitants and livestock into the ocean.

Not all the jökulhlaups are catastrophic, but when one occurs it is the subject of intense scientific interest. "We are now having a jökulhlaup from Grímsvötn," a member of the Science Institute of the University of Iceland wrote to a colleague in the U. S. Geological Survey, Washington, D.C., in March 1972. He continued that it was expected to culminate within a week. "It has been decided to build a road . . . and our road engineers are therefore very interested in this jökulhlaup. I myself have observed most of them since 1934. Our annual visit to Grímsvötn and studies of their behavior since 1958 have now suddenly become very important as we can foretell . . . their approximate maximum runoff. . . ."

This knowledge could be more than academic in value. The Vatnajökull icecap, Iceland's largest, has had a jökulhlaup almost every five years, in most cases the result of a subglacial volcanic eruption which did not penetrate through the icecap completely. The volume of water melted has in some cases amounted to the runoff of the Amazon River in the course of a few days.

A Highly Civilized Crater

(The Pozzuoli Uplift: Italy, 1970)

A justifiable regard for his own skin, the security of his family, and the protection of his home has made man generally prudent in his choice of living space. There is little elbowing for acreage on the brink of precipices, and competition for building sites in the paths of creeping glaciers is not keen.

Yet there are exceptions, which tend to become obscured by the acceptance of recurrent disaster as a condition of life. The population, and even the overpopulation, of high-risk real estate has been known in the past and continues to the present day. We have come to absorb as commonplace news of great floods that perennially make refugees out of thousands, of frequent tornadoes deadly to humans and their habitations in the Mississippi Basin, and of ruinous droughts that periodically afflict the Great Plains and southwestern United States.

Sheer economic necessity, the lack of a viable alternative, or the illusion of safety in numbers combine to outweigh practical considerations for the millions who inhabit areas where nature plays dice with their futures. For almost 600 miles of its 780-mile length, the Golden State of California is flawed by a north-south rift known as the San Andreas Fault. Geologists estimate that this "master fault" of an extensive system of geological flaws crisscrossing the state has slipped horizontally along its opposing surfaces at least 125 miles, perhaps as many as 350 or more, in the past 30 million years. In 1906, when the San Andreas Fault slipped a mere 21 feet, the San Francisco Earthquake resulted, causing the loss of 700 lives and $400 million in damage. On February 9, 1971, when a slip occurred along the smaller, transverse San Fernando Fault, 64 people (45 of them in a veterans' hospital) were killed, and total damage reached $1 billion. It was fortunate that the quake took place at 6:00 A.M., local time, before the start of the business day, or the effects might have been much more serious. And if the Lower Van Norman Lake

Dam, which the tremors brought close to collapse, had actually failed, this moderate-intensity earthquake could have been the worst in United States annals.

A report of the U. S. Geological Survey stated:

"Clearly, much of the damage could have been prevented had these zones been delineated, and zoned for open space or parks, before the San Fernando-Sylmar area was urbanized. It is encouraging for the future that many investigators feel confident that, armed with the knowledge gained at San Fernando, they can identify and delineate other similarly active thrust faults in this region and elsewhere in the Western United States."

It remains to be seen just how gratefully this kind of counsel will be received. The prospect that the investing and buying public in California or anywhere else will enthusiastically welcome the creation of greenbelts and woodland rambles on expanses of land begging for high rises and shopping centers seems somewhat remote.

California, however, will have to stand in line when the awards are conferred for the longest continuous display of *dolce far niente* in the face of clear indications to the contrary. The laurel wreath will unquestionably go to Pozzuoli, on Italy's west coast, which was literally founded on the lid of a volcano and has been flourishing there, albeit a bit unsteadily, ever since.

"And we came the next day to Puteoli." This passage appears in the Book of Acts (28:13), as written by one of the Apostle Paul's companions on his fateful journey to Rome in A.D. 61. The port at which they landed was the Pozzuoli of modern times. It was ancient even then, dating at least from 520 B.C., when it was established by a colony of Sabines from nearby Cumae.

The character of the area and the dangers it embodied must have been apparent to the first settlers, whoever they may have been. The landscape is pitted with volcanic craters from a few yards to a half mile in width; there are fumaroles emitting jets of steam, pits of boiling mud, sulfur springs, and blocks and fragments of volcanic material strewn thickly about. As it often does, myth bears witness to reality, and Ulysses is said to have sailed into Puteoli in search of the portals of Hades. He found them here and descended into the underworld to consult with spirits who could direct him back to his homeland.

Nonetheless, it is possible to appreciate why Pozzuolians through the centuries have not been discouraged by the evidence of their eyes and senses into abandoning their city. It is situated on the northern horn of the spectacular 35-mile crescent of the Bay of Naples. Separated on the east from the main portion of the bay by the promontory of Posilopo, it is sheltered on the seaward side by a headland known as Cape Miseno. Its harbor opens out toward the south, parallel with the coast line, and in the days of sail provided easy access to ships that caught the fair south wind on the final leg of their journey—as did St. Paul's on passing through the Strait of Messina.

In the first century B.C., Puteoli grew to be the most important commercial port in Italy. Much of the corn supply of Rome came through here, continuing to its destination 125 miles north along roads connecting with the Appian Way. Seneca describes the joy of the residents of the city on the arrival of the grain fleet. His contemporary, the poet Statius, tells of the libations to the goddess Minerva made by the crew of the first ship to arrive as it passed the temple that bore her name on the cliffs of Sorrento, which (then as now) stands opposite Puteoli at the lower end of the Bay of Naples.

Puteoli subsided into relative obscurity as Naples rose in influence and supplanted it as the queen city of the Tyrrhenian Sea. But the ancient harbor never lost its charm, which was complemented inland by the beauty of the Campanian Plain. This great district passed successively through Greek, Etruscan, Samnite, and Roman hands to emerge as a province of modern Italy celebrated for the mildness of its climate and the abundance and variety of its fruits and vegetables. It is incredibly rich, as well, in legend and history, much of which is centered in Puteoli and its immediate surrounding area, which was named by the Greeks the Campi Phlegraei and is the best, most fertile portion of the Campania.

Bright sun, clear skies, blue water, and a lush countryside—all these combined as they do in every age to attract seekers after pleasure and relaxation. The patricians of Rome made Puteoli and its environs their playground, drawn there in part by the thermal baths and mineral springs. Cicero had a country house in Puteoli itself and another on the edge of the adjoining Lucrine Lake. Julius Caesar, Pompey, and Nero all had villas in neighboring Baiae, a resort which achieved a lurid reputation for the quality of both its residents and its entertainment.

Among the local events was the construction by the Emperor

Caligula, another property owner, of a bridge of boats across the Lucrine Lake to Puteoli, one of the excesses that led to his eventual assassination. Nero achieved even greater infamy by contriving the death of his mother, Agrippina, in her villa on Lake Lucrine, which was not a healthy spot for members of the imperial family. Nero so feared the consequences of his deed that he spent a six-month cooling-off period away from Rome at his Baiae property. During this time he was cheered by numerous messages of congratulation from Campanian supporters. Reassured that his act was regarded as one of statesmanship rather than outright murder, he finally returned to the capital with a clear conscience.

The political turbulence of the times was echoed by some disturbing natural occurrences involving Vesuvius, which rose ornamentally above the Bay of Naples only a dozen miles southeast of Puteoli. The mountain had not been in eruption since the Greek settlement of this part of Italy, approximately eight hundred years before, and its threat had faded from living memory. Although the volcanic composition of the cone had been detected by the geographer Strabo, who had also pointed out the fire-eaten nature of its rocks, the slopes were intensively cultivated, and towns lay close to its foot. Among the larger were Herculaneum at the western base and Pompeii at the southeastern.

In 63 B.C., Vesuvius began to exhibit awakened volcanic energies in the form of a series of violent earthquakes that continued intermittently for sixteen years. The first of these inflicted terrible damage on both Herculaneum and Pompeii, destroying most of their public buildings. The citizens of the two towns immediately fell to the task of restoration. The work was still going on in A.D. 79, when on August 24 the earthquakes, which had been growing more violent, culminated in the tremendous, classic eruption of Vesuvius that buried not only Pompeii and Herculaneum but also Stabiae, a resort at the southern extremity of the Bay of Naples. Stabiae was rebuilt at a slightly removed point and is modern Castellammare di Stabia; Pompeii is famous for the excavation of its buildings; but the same mixture of volcanic ash, stones, and cinders that fell on Herculaneum was drenched in water and hardened to encase the city in a concretelike tuff anywhere from 18 to 65 feet deep.

From then on, Vesuvius never fully subsided. It has burst out again into major eruptions over the centuries, and others are anticipated for the future. A new age of anxiety began for Puteoli, and with it the tradition that, when Vesuvius is at its quietest, the Campi Phlegraei is at its most active volcanically, and vice versa.

Even with ruined Pompeii, Herculaneum, and Stabiae before them as an object lesson, with Vesuvius in view as an inescapable reminder, and with the ground underfoot giving a tentative wriggle now and then, the citizens of Puteoli did not act as if their own days were numbered. In fact, preparations for bigger and better civic attractions continued, and there is strong evidence that the amphitheater, whose stage arrangements exist today in a remarkable state of preservation, dates from the very years when Puteoli's future may have seemed uncertain.

It is known that the older of two amphitheaters built in Puteoli lasted at least until A.D. 66, since this was the year that the ubiquitous Nero himself fought there ceremonially in gladiatorial contests held in honor of a visiting vassal king. The second amphitheater had interesting facilities for flooding the arena, presumably to stage mock sea battles. These, however, were superseded in turn by subterranean corridors and cages by which animals or gladiators, or both, could be instantly introduced through trap doors to any part of the games area.

The size of the project, and the subsequent improvements, testified to the determination of the city fathers to compete with the first-run houses in this particular form of Roman summer stock. The over-all dimensions of the structure were 482 by 382 feet, as compared with the 615 by 510 feet of the Colosseum, largest in the Empire. Judging by the Colosseum's seating capacity of 50,000, the second Puteoli amphitheater must have held between 25,000 and 30,000 spectators. Inscriptions on the remains of its stone walls indicate that it was erected during the reign of one of the Flavian emperors who, beginning with Vespasian, ruled between A.D. 69 and 96. This means that construction began well after the Vesuvian earthquakes of the Christian Era started, and may have extended past the eruption that destroyed Pompeii. In any event, it was business as usual for Puteoli while its sister cities were being demolished twice over.

There have, however, been some close calls throughout a grace period which is now going on two thousand years. In the shadow of Naples, Puteoli has slid into a pleasant obscurity, thriving modestly and surviving into the twentieth century as the city of Pozzuoli, 1972 population 65,000. The Campi Phlegraei, now better known as the Terra di Lavoro, has escaped damage from repeated eruptions of Vesuvius but, in off years, has had some paroxysms of its own.

To the northeast of Pozzuoli is the Solfatara crater, a semiactive volcano known to the Romans as the Forum of Vulcan. It exhales sulfurous vapors and gives vent to liquid mud and hot mineral springs. In 1198 the Solfatara temporarily forsook its role as a regional curio and source of medicinal waters, with an eruption of lava that did substantial, but not irreparable, damage to Pozzuoli. It then reverted to its previous state and has achieved a fame beyond its geographical bounds by lending the name "solfatara" to any volcano which has ceased to erupt with lava or other solids and is content to emit noxious fumes.

In 1538, the shore line between Pozzuoli and Baiae suddenly began to bulge and hunch until it had rapidly reached a height of 20 feet above its normal level. After this initial convulsion the earth exploded into a shower of lava and volcanic fragments, forming a cone 456 feet high and 1.5 miles in circumference. This was the creation of Monte Nuovo, which took place in the course of a few days—some accounts maintain it was in a single night.

In its birth pangs, Monte Nuovo reduced the size of the Lucrine Lake, destroying a village and the remnants of Cicero's villa on the east bank. The effects on Baiae were academic: the lively and beautiful resort town had been abandoned for all time in 1500 because of a malarial scourge.

The uplifted earth gradually subsided, taking centuries to do so but leaving Monte Nuovo behind like a scaled-down Fuji. In a belated footnote to the events of 1538, almost every fish in the Lucrine Lake was killed in 1922 by exhalations of steam and hydrogen sulfide from vents on the lacerated east bank.

As if these upheavals in its own back yard were not sufficient, there were ominous portents from the seaward side as well. The island of Ischia, lying off the coast seven miles southeast of Pozzuoli, is part of the same volcanic system as the mainland. It has a continual history of eruptions from prehistoric times on, and was deserted by its original population about 500 B.C. for that reason. The last eruption was in 1301, but in 1881 and again in 1883 Ischia experienced severe earthquakes, of which the second took 1,700 lives.

Taking stock after the Ischia quakes, Pozzuoli could easily see itself as lying within a circle of geological disaster. The question was not so much whether, but when the circle would narrow sufficiently to make the city itself the focus of events which had so far struck around it at every point of the compass.

It did not happen for almost another hundred years, until 1970. On March 4 of that year Pozzuolians counted ten shocks off their coast. In the next two days the land on which their city stands rose steadily at least eight inches, cracking some building walls and leaving the docking area too shallow for use.

Only the children were made happy by the uplift, since the waters receded from the sea wall to leave a fine new swimming beach where none had existed before. Their elders were less delighted by this bonus. In response to government warnings, 30,000 people fled Pozzuoli, spurred by the image of Pompeii and other lost cities.

Teams of scientists from research centers in Italy and France converged on Pozzuoli. They took the temperature of the Solfatara, analyzed the emissions of the fumaroles, and ringed the city with instrumentation that monitored it as continuously as a patient under intensive care.

The crater was found to be the same in heat level as when it was measured in 1967, 1933, and 1907. Analysis of the gases emitted showed them to be of stable composition with no changes. No evidence was found of the shocks reported off the coast. Not a single seismic event was recorded by a network of seismographs installed for a ten-day period (March 6–15) by the geologists of the Laboratoire International de Recherches Volcanologiques.

And still the uplift, known as a bradyseism, continued. From March 20 through April 20, the land rose another two and a half inches over an area roughly 16 miles from east to west and 6 miles from north to south. No new eruptive fractures appeared; the scientists were definitely of the opinion that, anomalous as the uprise might be, it was taking place too gradually to be used as a pretext for evacuation by itself alone.

A great many residents who had departed took this reassuring word at face value. All but 10,000 returned by the last week in March, to be greeted by a series of small earthquakes. Typical of these was one on May 3, which registered an intensity of 3 on the Mercalli scale. Absolutely no other unusual phenomena could be detected, by either seismology, geochemistry, geomagnetism, heat or temperature measurements.

The absence of data on which to base definitive judgments is characteristic of the aura of mystery surrounding the Pozzuoli Uplift. Scientists are not agreed on the depth at which the actual subterranean movements take place that are responsible for the elevation of the land surface. Some say the focus is only a mile or two in depth; others estimate it to be ten or twenty times deeper—possibly more.

There seems, also, to be a gap in the early chronicles of Pozzuoli's ups and downs. Temple porticoes along the shore have been drowned by the sea as the result of earth movements. The Roman market, dating from the first century A.D. and erroneously called the Temple of Serapis because of its statue of that Egypto-Grecian god, is semisubmerged. But its ancient pillars are eaten away by shellfish activity to a height of 15 feet above the ground line. If the evidence is to be credited, it points to some period during the Christian Era when the market subsided beneath the sea and remained there for a substantial length of time, a fit subject for legends about sunken cities. How many times Pozzuoli or sections of it have gone through this cycle can only be guessed.

Looking ahead, the Italian government has set a watch on Pozzuoli, as it has watched Vesuvius since 1844. They are confident that their scientists and scientific instrumentation will be able to give warning of any eruption far enough in advance to permit evacuation.

Japan is another country where bradyseisms are peculiar to certain areas of seismic and volcanic activity. Japanese scientists are among those who have been studying the various Pozzuolian phenomena and their interrelationships in the hope of improving their ability to predict earthquakes and eruptions.

In another age they might have consulted the Cumaean Sibyl. She had her cave here, and it is still shown to those who are skeptical. Unfortunately, she is no longer a resident, owing to an oversight which even her powers did not catch. Apollo loved her and offered her as many years of life as she could hold grains of sand in her hand. She accepted the gift but forgot to ask for eternal youth as well, and so finally withered away into dust, leaving the foretelling of the future to mortals.

A Deferred Disaster Comes Due

(The Tuscania Earthquake: Italy, 1970)

Antiquity is no permanent charm against the laws of probability, as the city of Tuscania discovered after twenty-three centuries. On a Saturday afternoon and evening, February 6, 1971, twin quakes struck the ancient town located about 50 miles north of Rome, on a plain of volcanic origin long known for its seismic capriciousness. At least 18 persons were killed and 270 injured, with all 7,000 residents left homeless. An aftershock the next day destroyed one of the few buildings left standing. It was the worst such disaster in Italy since earthquakes in western Sicily three years previously killed 316 people and left another 90,000 without a place to live.

The earthquakes were felt as far away as Rome and Trieste. Some damage was reported in nearby towns, with that in Tuscania alone estimated at $41.6 million. The public works director of the region gave out approximate figures of 100 buildings destroyed and 400 damaged in the town's medieval center. Those structures that were not shattered were so weakened that their occupants were forced to leave.

In addition to the human toll, there was the grievous prospect that the damage suffered by rare and valuable Etruscan art—much of it preserved in pre-Christian tombs—might be as serious as that to the priceless Renaissance art treasures of Florence during the 1966 floods.

The earthquake, which experts said was probably of volcanic origin, was not exceptionally strong. It measured approximately 6 on the 12-point Mercalli scale and only 5 on the Richter scale.[1] The contrast

[1] The Richter scale was devised in 1935 by Charles F. Richter, an American seismologist at the California Institute of Technology. It gives each earthquake a numerical rating which is the index of the quake's energy at its source. This index number is called the earthquake's magnitude.

An earthquake of magnitude 2.5 has an energy of less than 10^{17} ergs, approximately the amount of energy released by the burning of 100 gallons of gasoline. Small tremors of this magnitude occur very frequently.

between these relatively low readings of magnitude and the extent of the damage points up the special nature of earthquakes in the Mediterranean and Middle Eastern areas: even small shocks there can have devastating effects. Historical examples are found in Yugoslavia and Morocco.

An earthquake of magnitude 4.5 (10^{27} ergs) can cause very local minor damage; one of magnitude 6 has a high potential for destruction.

Each whole number on the Richter scale, which is logarithmic, represents a shock ten times greater than the previous whole number. In developing his scale, Richter assigned an index of magnitude for every earthquake up until then which had been measured by instrumentation. Before 1972, the most destructive earthquake recorded had a magnitude of 8.6. In May of that year an earthquake in Chile (not detailed in this book) broke the record by measuring 9–9.2.

The Ice Falls of Huascarán

(The Burial of Yungay City: Peru, 1970)

It has been called the Switzerland of South America, but the Callejón de Huaylas needs no picture postcard evocations of middle Europe to do it justice. It is a natural wonder in its own right, a spectacularly beautiful valley carved deep in a double spur of the Cordillera Occidental, the continental divide of the Peruvian Andes.

The western wall of the valley, known as the Cordillera Negra for its black volcanic rock, rises out of the coastal plain less than 30 miles from the Pacific Ocean. It parallels the shore for 100 miles, a forbidding escarpment whose ridge line undulates between 14,000 and 16,000 feet in height with hardly a practicable pass in its entire length.

To the east, the valley is bounded by the 70-mile-long Cordillera Blanca which, unlike its bare and gloomy companion range, is broken by a succession of jagged peaks high enough to be perpetually capped in the gleaming ice and snow that give it its name. Centered in the Cordillera Blanca, and dominating the whole of the Callejón de Huaylas, is Mount Huascarán.

Huascarán's twin towers, the northernmost topping 21,000 feet and the southern 22,000, are the highest in Peru, dwarfing Europe's Mount Blanc and the Matterhorn by a vertical mile. In all of South America, only the extinct volcano Aconcagua, on the Chile-Argentina border, is higher than Huascarán—and that by a mere 280 feet.

The valley itself is narrow (*callejón* literally translates as "corridor" or "alley"), averaging 25 miles in width. It is one of the most densely populated parts of the region. Good soil, plentiful water, and a favorable climate combine to produce a variety of crops, tropical in the lower reaches and temperate in the upper. Prize cattle and fat sheep graze the slopes, and the rich agricultural heritage comes down from before the days of the Inca Empire, under which it reached its height. It was here that the Spaniards first found the potato, which had been bred by the Indians of Peru from the bitter tubers native to the continent.

Flowing through the lush floor of the valley in a deep canyon is the largest of Peru's coastal rivers, the Río Santa. It rises east of the continental divide, enters the valley, and follows a northerly course. At the upper end of the Callejón de Huaylas the Santa makes a dramatic turn to the west before plunging into the Pacific through one of the world's great cataracts. In the rocky gorge at this final bend of the river is a hydroelectric plant which powers the steel mills of Chimbote, a developing industrial and port city lying directly below.

But as it threads longitudinally through the valley, the Santa belongs to another world. It widens out intermittently into a series of basins, each of which became the setting for a town, strung one after the other along the banks like ornamental beads on a silver filament.

At the southern head of the valley came Recuay, followed by the provincial capital, Huarás, then Anta, Carhuás, Yungay, with its handsome plaza, next Carás and Huaylas, from which the Callejón takes its name, and farthest downstream, Huallanca. Bearing out the promise of the colorful names was an endless variety of architectural fantasy and invention; brightly tiled roofs set off by bone-white balconies almost touching across narrow pavements; doorways painted in pastel shades of turquoise blue on one side of a street and umber or green on the other; gaily colored walls of brick or adobe decorated with bas-reliefs and other forms of Peruvian folk art.

Travelers—they could hardly be called tourists—spoke of the Callejón de Huaylas with that special regard for one of the world's unique and unspoiled places. For picturesqueness, Huarás rivaled Cuzco, the ancient Inca capital, and was more intimate and less self-conscious in the bargain. Its famed mineral springs fed such hedonistic delights as hot sulfur swimming pools, and its museum contained local pre-Columbian art that the archaeologists had not even as yet classified. Nearby, in the double Llanganuco lakes, 14,000 feet high in the Cordillera Blanca, was some of the best trout fishing in the world.

The Plaza de Armas in central Yungay was framed in royal palms and planted with beds of white roses and crimson gladioli. Its fountain threw gentle jets of crystalline water into the clear mountain air, near a granite statue of a dignified mustachioed general from Bolivar's time. The church was of plain brick but paid tribute inside to the fullness of life around it with a ceiling intricately painted in floral designs, culminating in a giant daisy where transept and nave converged.

Throughout the Callejón de Huaylas, whether in the towns or the countryside, there was always something to which the eye was attracted because of native handiwork or natural beauty. It was difficult to conceive of a more unlikely site for a great catastrophe. Yet the valley became just that in the backlash of one of the most destructive earthquakes in the history of the Western Hemisphere.

Peru is part of the earthquake belt that rims the land masses bordering on the Pacific Ocean in an arc extending from the tip of Chile northward to Alaska, and then south through the islands of Japan. The South American segment of this arc, 4,000 miles long, is shaken by earth tremors and endowed with active volcanoes, oceanic trenches, and young, still rising mountain chains. Scientists believe that these manifestations of "geological immaturity" result from a progressive underthrusting of the Pacific Ocean beneath the continental shelf.

Although numerous large earthquakes are on record over the past century in the Peruvian sector, the Santa River Valley has been relatively untouched. Residents of the valley could not recall an earthquake of major proportions in their area for at least three generations.

The potential for death and destruction was certainly there, however. The Callejón de Huaylas must have had its tragic share in the earthquake of February 4, 1797, that killed as many as 40,000 people between Cuzco in the south and Quito, more than 1,000 miles to the north. And in the quakes of August 13–15, 1868, that killed another 25,000 in Peru and Ecuador.

For the rest, the valley was unscathed by the lesser Peruvian quakes which took place between 1940 and 1966. These were located principally south of the Callejón, in the Lima and Arequipa areas, exacting scattered tolls of dead and injured.

Immune—almost—from earthquakes in recent times, the valley paid for its physical charms in another coin—the avalanche. Mud slides, rock showers, and every conceivable kind of material cascading from the mountainsides that hemmed it have plagued the Callejón since the dawn of its recorded history. The chief villain is the Cordillera Blanca, whose upper slopes are so steep that grades of 45 to 90 degrees are the rule rather than the exception. In some areas, small icefalls take place several times a day; in others, major avalanches are reported every few months or years.

The deep U-shaped glacial valleys in both flanks of the Cordillera Blanca drain its slopes, those on its western side into the Río Santa. The glacial moraine, consisting of rock, pebble, shale, and other sediment, has formed lakes, some of which gradually cut through their morainal dams and disappeared. Others have suddenly burst their dams, owing to the overtopping of the morainal wall by waves generated by ice- or rockfalls. When these lakes suddenly drain in this manner, catastrophic mudflows, called *huaycos,* can result, such as the one which destroyed one quarter of Huarás in 1941. Atop the 13,000-foot pass leading across the Cordillera Negra from Casma on the coast, it is possible to look down on Huarás, which itself lies at an elevation of 9,800 feet, and see an expanse of rounded boulders on the edge of the city. They shine white in the sun, but their beauty is deceptive. They are the projectiles that flew down the mountainside in the night to entomb 6,000 sleeping residents and remained to mark the place where they died.

The Cordillera Blanca, and Huascarán in particular, is a smiling mass murderer that wipes out entire communities. The unfortunate village of Ranrahirca, which adjoined the city of Yungay, has been repeatedly singled out for annihilation. In the early evening of January 10, 1962, an estimated 3 million tons of Huascarán's northern peak tore away from the face of the mountain. The huge rock mass bounced five times in the river gorge and then, slowing down to 60 miles per hour, crushed all but one of Ranrahirca's houses and buried 3,500 villagers. The rain of boulders traveled a distance of 9 miles in seven minutes, dropping 13,000 feet.

The survivors of the rockfall, and of the epidemic of typhus that followed, rebuilt Ranrahirca, putting up houses of adobe tile or cement to replace the earlier structures of cane. A visitor to the restored village observed that the new buildings looked solid, but puny against the mountain. It was a prophetic remark.

The remains of many of the Ranrahirca victims were deposited in the wall niches of the Yungay cemetery. In gratitude for having escaped the martyrdom of their neighbors, the residents of Yungay put a figure of Christ with arms outspread in benediction over their cemetery—similar in form and spirit to the Christ of Coronado at Rio de Janeiro.

In the seasons following the Ranrahirca disaster, Peru underwent a political change with the accession of a military government. Among the policies pressed forward by the revolutionary regime were the industrial development of the country and measures to alleviate the condition of the Indians, who comprised half the population of 13.5 million. Since the advent of Pizarro in 1425 and their subjugation to Spain, they had been exploited by a Spanish elite as they had once been ruled by an Inca elite. Now, still speaking the Quechua dialect predominating in the vanished Indian empire, they dwelt for the most part isolated from contact with the authorities. In remote Andean communities, they lived and died anonymously, their homes, adobe huts, often collapsing in wind and rainstorms. Their growing numbers, however, were making them more obtrusive in the cities and of greater social concern generally, and the worried government was looking forward to a national census in 1971 which would, for the first time, give some statistical proportion to the problem.

But before these plans could materialize, natural forces intervened which would change the final count. Later, geologists and geophysicists again alluded to the immaturity of the Andes. The youthful mountains, it was said, displace earth as they rise, pushing it down into faults existing deep in the sea bed off the coast. The result is a series of surface movements commonly called earthquakes.

Despite the frequency of such earth movements, few were important enough to induce apprehension among Peru's millions. The last major quake had occurred in October 1966, when 110 died in the Lima-Callao area.

On May 31, 1970, there was a violent awakening for a 5,000-square-mile area of Peru and for the country as a whole. At 3:23 P.M., Eastern Standard Time, a shock with the magnitude of 7.9 on the Richter scale took place fifteen miles west of the port city of Chimbote, focused 35 miles under the sea bed. It was followed by at least thirty-seven aftershocks throughout the week that followed, distributed over an ocean area extending about 85 miles parallel with the coast and 40 miles out to sea.

That the motion originated in deep faults was borne out by the fact that there were no tsunamis, or seismic sea waves, as might have occurred if the ocean floor itself were wrenched and deformed. Nor was there any movement of known surface faults in the region or warping along the coast.

Various reports disagree slightly on whether the maximum intensity of the shocks inland, especially in the Callejón de Huaylas, was somewhat more or less than 7.9. The question is academic. Within roughly a 100-mile radius of Chimbote, unimaginable disaster struck the coastal and mountain regions with the capriciousness that characterizes nature on a blind rampage.

Most accounts described the tremors as having a pronounced side-to-side motion that made it difficult or impossible to walk or run, but not strong enough to throw people to the ground. But if humans could stand against the shocks, the adobe structures with which the area abounded could not, and it took fifteen seconds at the most before they began to collapse.

There is an uncanny parallel, suitably modified by the passage of more than two hundred years, between the Peru earthquake and the one that ruined Lisbon on November 1, 1755. The Lisbon quake also had its origin in the sea bed west of the city, and the three great shocks that demolished most of the houses in the lower part of the town killed 50,000 people in a few minutes. The number of victims was higher than it might have been, since it was All Saints Day and the churches were filled with worshippers.

Peru's earthquake took place on a Sunday afternoon, hours after church services had ended, but this was no ordinary Sunday. Thousands of devout Peruvian sports fans were indoors, listening to the broadcast from Mexico City of the World Soccer Matches between the Soviet Union and Mexico. The outcome was important, since Peru was in the finals. Only the fact that the quake began forty minutes before the traditional 4:00 P.M. Sunday film matinees saved many more from being trapped en masse in the theaters.

In the coastal region, with its population of 200,000, a mere 1,000 died. Of these, about 700 were in Chimbote, where 70 to 80 per cent of the buildings were destroyed or seriously damaged. Another 70 victims were Casma's, a miraculously low figure for a city of several thousand, which was totally leveled. The remainder were in such scattered towns as Huarmey, where destruction comparable to Chimbote's occurred and—farther to the north—Trujillo. This city, founded by Pizarro, and named by him after his birthplace in Spain, got off lightly with only 10 to 20 per cent of its structures damaged.

It was a different story inland, where time finally ran out for the Callejón de Huaylas. The clock in the cathedral church of Huarás stopped at 3:27 P.M., and doomsday began.

In the older part of the city, the red-tiled roofs of the two- and three-story buildings came down on their occupants, and the walls of adobe brick fell on those who fled down the narrow streets. In forty-five seconds, nine out of ten buildings were destroyed, and there were

800 known dead, with 16,000 more out of a population of 65,000 buried in the ruins. Two young women, Peace Corps members from the United States, were among them, struck down less than fifteen feet from their own doorway. An area of 100 square blocks was turned into a mass of rubble and broken walls, as much as 10 feet in depth.

Buildings of brick, with reinforced concrete walls, withstood the shocks with only slight damage. But even this type of construction did not avail where there were inadequate shear walls. One such building, a four-story school, collapsed as the pupils were performing a play, and the bodies of two small girls were found beneath that of the priest who had vainly tried to shield them.

What happened in Huarás was being repeated on a smaller scale in dozens of communities throughout the Callejón de Huaylas and hundreds of remote spots on the slopes of the mountain ranges that hemmed it in. Mancas, situated on the western heights of the Cordillera Blanca above the Santa River, was typical. Roughly 70 per cent of its adobe buildings were destroyed and 3,100 of its 5,000 inhabitants killed. In less than a minute the Peru earthquake secured itself a prominent place on the list of the world's natural holocausts.

It was unimaginable that anything could overshadow the devastation that had already taken place, but the unimaginable happened.

A slab of ice and rock a half mile wide detached itself from the western face of Huascarán's northern peak and began its descent toward the Río Santa 18,000 feet below.

The original mass that broke loose is estimated at considerably more than 35 million cubic feet. As it moved at incredible speed down the 70-degree slope of Huascarán, it began to generate awesome properties in response to the natural laws invoked by its huge size and velocity. In its slide, the avalanche degenerated into a highly liquid mudflow; part of the ice melted through friction and other energy release, and water and water-saturated sediments were picked up in the Llanganuco Valley.

Accompanied by a deafening noise, and preceded by a strong turbulence, the fluid mass is thought to have created its own cushion by entrapping and compressing air beneath it. In this way it was able to override obstructions that might have slowed or stopped it, and even ridges of loose morainal material were later found to be undisturbed in its wake.

Hurling 3-ton boulders half a mile, burying *en passant* a party of 15 mountaineers encamped between the two lakes of the Llanganuco, wreaking havoc on people, livestock, vegetation, and buildings in the valley of the same name, the avalanche gave a foretaste of what was in store for Yungay and Ranrahirca directly below.

The Llanganuco Valley, cut transversely into the western flank of Huascarán, intersects the Río Santa at almost right angles. Now the Llanganuco acted as a chute for the immense mass which, traveling at various speeds between 135 and 270 miles per hour as its pathway dipped and eased, ate up the 9 miles to the edge of the Santa River in less than four minutes. At the 8,500-foot level, it forked into three tongues, or lobes, encompassing about 6 square miles. The first lobe buried Ranrahirca and all but 50 of its 1,800 residents; the second obliterated several smaller communities. The third hurtled toward Yungay, where a ridge 300 to 600 feet high stood between the city and the mouth of the Llanganuco Valley. It had acted as protection against landslides in the past.

Not this time. Before the eyes of the few horrified onlookers who lived to tell what they saw, a wall of debris as high as a ten-story building reared up over the barrier and, like an incarnation of the apocalyptic beast, fell on Yungay and swallowed the city whole.

In that instant, more than 15,000 of Yungay's 19,000 citizens died, together with unknown thousands of Sunday visitors and vendors of farm produce there for the market day. The city lay buried 17 feet deep, instantly transformed into a modern Pompeii. Of the Plaza de Armas nothing remained but the tops of four palm trees. Whatever the great slide could not cover, it toppled; the church with a belfry tower 36 feet high disappeared.

The advancing wall of debris carried across the Río Santa and leaped the opposite bank, in some places as much as 150 feet vertically. It overran the other side of the river for several hundred yards more, destroying much of a small village in the process. A macabre parody of the Day of Judgment ensued: the Santa River ran backward for a mile or more as millions of tons of rocks and ice disrupted its course, and the avalanche—having entombed the living of Yungay—spared its cemetery but ripped open the ossuary and resurrected the bones of the dead. The statue of Christ still stood, arms outspread as before, but now in compassion rather than benediction.

The southern lobe of the avalanche continued down the Callejón de Huaylas at a speed of 40 miles per hour, reaching the edge of Carás in fifteen minutes and partially inundating the airport at the southern edge of the town. A football game was scheduled for 4:00 P.M. between the teams of Carás and Huallanca, its neighbor to the north.

"We were warming up when the earthquake began, and we were still talking about it on the field when the sea of mud came at us," one Huallanca player told a reporter. "It stopped just short of the opponents' goal."

Elsewhere, innumerable other landslides did not stop fortuitously but contributed to the toll of dead and injured. Flooding, always a real danger, was aggravated by the action of both the main and subsidiary slides in forming an artificial lake behind the Río Santa and also damming the upper of the two Llanganuco lakes so that its level rose perilously. Fissuring and soil compaction took place the length and breadth of the valley; the mud left by the Huascarán avalanche, ranging in depth from an inch to 30 feet in places, rapidly dried to the hardness of brick.

Huge boulders lay about, some the dead weight of a full-sized seagoing freighter. The largest, 14,000 tons, was left at the site of Ranrahirca. A 7,000-ton giant marked the Yungay lobe. Most of the rock fragments were angular and fresh, as though torn from the outcrop. There were, however, many from previous landslides which had apparently been picked up and carried along by the Huascarán avalanche.

The total volume of solid material moved could easily have been on the order of 65 million cubic yards, according to estimates made by geologists on the scene a few days later. This is more than one quarter of the 240 million cubic yards excavated during the building of the Panama Canal, a project which took ten years.

The violent displacement and redeposit of immense masses of material with such suddenness had incalculable effects within the confines of the valley. In following the Santa's course, the avalanche destroyed bridges, roads, communications facilities, fields, and power lines. But the greatest damage was human, and the Callejón de Huaylas lay like an open wound.

Peruvian Air Force pilots reported that visibility was hampered by clouds of dust from continuing avalanches which were mingling with the seasonal fogs of the Andes. There was, however, sufficient visibility to piece together a picture of survivors clinging to the mountainsides (pitifully few at Yungay and Ranrahirca), valleys which had become raging rivers fed by burst dams, roads hidden under tons of rock, and scores of communities literally wiped off the face of the earth.

The disaster had been so precipitous, so complete, that it was days before its full impact was grasped. First accounts spoke of 200 casualties but climbed inexorably toward the thousands as additional stories were filed and government spokesmen, reluctant to admit the worst, began cautiously to mention increasingly greater numbers, hazarding the outside possibility of 30,000 deaths.

It was to be many more. The true figures were delayed for weeks because of disrupted communications and ground routes blocked by landslides. They were finally released at the beginning of July 1970 by Peru's National Emergency Committee: 70,000 lives lost; 50,000 injured; roughly 186,000 buildings, or 80 per cent of those in the afflicted area, destroyed or rendered uninhabitable.

There were also 200,000 homeless, and as news of relief supplies filtered in, these rural Peruvians began an exodus to the cities out of their devastated Andean highlands which were now gripped by the chill of the subequatorial autumn.

Some, however, could not be enticed out of their mountain fastness even by cold, hunger, or medical need. Three weeks after the catastrophe many Indians were still hiding their injured and orphaned out of a distrust for Spanish-speaking people which has persisted for four centuries. A great number of the survivors were also suffering from a lack of alcohol and narcotics, long used by Indians, not only to deaden the privation and pain which is their daily lot, but to stimulate themselves to perform prodigies of labor besides.

An even deeper reason than traditional fear prevented the Indians from permitting medical workers and aid teams to help them. Their suspicion of outsiders is based in large part on the common practice of obtaining Indian children by ruse in rural areas and raising them as servants. Among many Peruvians of European stock, the Indian is considered something less than human, and his use as a beast of burden on plantations has not completely died out. Roman Catholic and Protestant missionaries active in the region doubted that anything, including government broadcasts guaranteeing protection for the orphans, would move the Indians to surrender them for care and treatment.

International relief efforts took on unseemly overtones as the great powers competed for recognition as humanitarians. The United States flew in shipments of medicine and food, one of them accompanied by

America's new First Lady, Mrs. Nixon, on a condolence call. The Soviet Union next established a dramatic airlift over the Great Circle route which lasted until one of its planes went down in the sea, and the others were diverted to search for it. Meanwhile, supplies piled up at ports of entry while emergency crews puzzled over how to get them to the interior.

Newspaper stories dwindled. Both the dead and the living faded from public awareness. In Peru, plans were under way to rebuild the ruined cities and towns, except for those whose sites invited similar disasters.

The decades during which it was spared a major earthquake were at best uneasy ones for the Santa River Valley, marked by landslides that sacrificed a village here, a town there. Then the pact, if such it was, ended in a double tragedy that enveloped all of the valley in both a quake and an avalanche within minutes of each other. On that day the Callejón de Huaylas earned itself another truce with the destructive forces of nature, unfortunately with no guarantee as to its length or the toll which would be exacted when it was again broken.

"It's the Water, Not the Wind, That Kills"

(The Bay of Bengal Storm Surge: East Pakistan, 1970)

There are disaster-prone parts of the world where human life is said to be held very cheap. This assertion is strongest among people who live at a safe distance and is confirmed on occasion by some firsthand observers who can always make the last train out by displaying the right passport. The native populations left behind to serve as victims are assumed to accept their collective role with an impassivity bred of resignation. The 900,000 Chinese who died in the Yellow River flood of 1887 were not consulted on this point; nor were the 5 million Indians who perished in Bombay and the southern provinces during the 1876–78 famine. Their huge numbers and the passage of time clothe them in an anonymity that makes it possible to contemplate their plight with an impersonal detachment. In enormities of this sort—whether perpetrated by man or the elements—statistics tend to triumph over sensitivities.

It may come as a shock to traditionalists, therefore, that several million illiterate and struggling peasants who have been decimated periodically by hurricanes throughout their history are rising up to reject their lot. Rather than accept these deadly windstorms as a species of God-sent population control, they lay their effects and the suffering they cause as much to social and political inequities as to aberrations of the upper atmosphere.

The setting for this rebellion against both nature and fate is the country which in 1972 transformed itself violently from East Pakistan into the new nation of Bangladesh. Except for the tea-growing mountains of Darjeeling far to the north and the Chittagong Hills to the east, the entire land is the butt of a practical joke of cosmic proportions. For most of its 55,000-square-mile area, it scarcely rises above sea level, stretching flat for hundreds of miles in every direction without an undulation.

The forces that shaped this unusual terrain are two mighty rivers of India, the Ganges and the Brahmaputra. Both originate in the Himalayas and, flowing more than 1,200 and 1,500 miles from the northwest and northeast respectively, meet in the center of this populous land bridge between the Indian and Burmese peninsulas. Joining into a single river known as the Padma, they roll on through a broad estuary into the Bay of Bengal, 220 miles south. Together with their tributaries, they have the force of seven Mississippis.

Wide as it is, the estuary is not the only pathway to the sea for the Ganges and the Brahmaputra. As the two great rivers enter the region they are slowed and checked by the dead level of the vast alluvial plain. They become sluggish; their main channels divide and subdivide, and each new stream throws off another set of distributaries, interlacing to form the delta of the Bengal. The silt is no longer carried along by the waters but is deposited, so that the beds of the streams gradually rise above the horizontal floor of the surrounding countryside. Basins, pools, and swamps spread everywhere, and narrow waterways vein the land.

In the northern angle of the Ganges-Brahmaputra Delta, the land lies under a blanket of alluvial soil which in places has been sounded to depths of almost 500 feet. Without underlying mineral resources—and with agriculture and inland fishing the sole industries—it is intensely cultivated. The monotony of the landscape is broken only by fresh greenery, lines of trees tracing the numerous watercourses and clumps of bamboo marking the villages.

In the southern coastal belt the same kind of soil has been built up into chars, or offshore islands, which emerge, change shape and size, and sometimes disappear under the influence of the yearly floods. Many of these are uninhabited; others, larger and more permanent, are the sites of villages containing perhaps one million of the 2 million Bengalis living at the mouth of the estuary and along the adjoining shores.

The fingers of alluvial soil invade the shallows of the Bay of Bengal on an east-west front of almost two hundred miles. Mingling haphazardly with the salt waters, the coast line is intricately convoluted and indented. The middle segment of this ill-defined no man's land is called the Mouths of the Ganges, after the silted-up, vegetation-choked indentations through which the river once found its way to the sea before cutting new channels. Part of the area is a trackless salt marsh known as the Sundarbans, for the forests composed of Sundri trees that grow there.

The mainland and the islands not only invite the cultivation of rice but demand it. No other crop will grow in sufficient quantity to feed an exploding population of 75 million in an area the size of Arkansas (pop. 2 million). Even in a good year the three harvests of this staple are not quite enough, and a single monsoon rain failure is enough to make the ever impending threat of famine a reality.

Water—from above and from below—is at one and the same time both the blessing and the curse of the land. The average mean rainfall is 82 inches per year—about the same as New York's—but so unevenly distributed that some sections receive nothing while others are deluged with 200 inches. The melting snows of the Himalayas swell the rivers just as the rains of the summer monsoons strike, and by June most of the country, where flood control is minimal, is inundated.

The rainy season usually stretches from October to December, when the jet stream pushes south from the Himalayas, preventing the rain-bearing air masses from penetrating the mountains and causing them to drench the lowlands instead.

As a result, the inhabitants of the delta live an almost amphibian existence, and life has been molded into some strange patterns. In the coastal region peasants commute daily by sailboat between their villages and the rice paddies on the chars. They carry with them as passengers the bullocks used for plowing. The waterways are crowded with this picturesque traffic, born of a shotgun marriage between land and sea.

There are other great river deltas which serve as the rice bowls for their countries and where the seasons follow one another in predictably routine fashion. The delta of the Bengal is not one of them. The millions of Bengalis who are tied by dire necessity to their coastal rice fields are staked out on the ancestral killing ground of a breed of hurricane that is spawned in the Indian Ocean and expends its murderous energies northward.

In this part of Asia hurricanes are known as cyclones (elsewhere they are called typhoons or baguios), and their effect on the land rimming the northern end of the Bay of Bengal is catastrophic. Winds originating northwest of Sumatra are funneled at speeds exceeding 100 miles per hour up the narrowing head of the bay, driving the sea before them.

The bay is so shallow that the heaped-up water cannot flow back as it does in areas where the sea floor shelves upward at a gradient which creates a deep countercurrent. The driven water surges over the flatlands in swift waves known as tidal bores. Ten feet high or more, they sweep everything in their path, each increment of height and velocity measured in thousands of additional deaths. Dikes built to cope with southwest monsoon tides four or five feet above normal can do little to check these tidal bores. In 1737 a storm surge killed 300,000 of the region's peasants. In October 1960 a storm surge left between 15,000 and 50,000 dead, the uncertainty in the exact number reflecting a political structure too loose to produce accurate figures. In that year the entire area, comprising two thirds of historic Bengal Province, was East Pakistan, as it still was in November 1970.

It was the month that headlines were proclaiming the death of Charles de Gaulle, and another military man who had become the chief of his state was being accorded more modest notice on the inside pages of leading Western newspapers. General Agha Mohammad Yahya Khan, President of Pakistan, was reported as receiving a warm welcome in Peking, last stop on a shopping trip for arms aid which had already taken him with indifferent success to Moscow and Washington.

After five days General Yahya returned home, preceded by a communiqué that his talks with the Chinese had culminated in an economic co-operation pact. The luster of his diplomatic triumph was dulled somewhat by the fact that uncounted thousands of his countrymen had perished miserably during his absence.

They had died in a Bay of Bengal storm surge that struck at midnight on Friday, November 13. General Yahya waited until November 15 to fly back from Peking, as previously planned, and another ten days before making a tour of the devastated area. These delays were later thrown up to him as the first visible evidence of an unconscionable lack of concern on the part of the government. For the victims were all East Pakistanis or, as they preferred to think of themselves, Bengalis. The compromise partition of 1947 had joined them with the other predominantly Moslem provinces in India's northwest as the nation of Pakistan. But they were separated from the Punjabis of West Pakistan by a 1,100-mile tongue of Indian territory and by cultural, racial, and language differences that not even a common religion could bridge.

Economic imbalance was the sorest point. With due allowance for census figures which lag behind a galloping birth rate of 3 per cent per year, there were an estimated 55 million people living in West Pakistan's 310,000 square miles, giving it a population density of 175 inhabitants per square mile. The comparable figures for East Pakistan were 75 million people crammed into 55,000 square miles, for a population density which was among the highest in the world: 1,350 per square mile. Aggravating the disparity was the difference in average annual income: $60 per person in East Pakistan as against twice that for the West. And the deepest grievance of all was that the politically dominant Punjabis of West Pakistan received the overwhelming proportion of development funds, factories, public works projects, and defense facilities. The backward region of East Pakistan remained mired in its rice economy with only a few scattered minor industries in addition.

The regime of General Yahya Khan was a military one, dating from March 1969, when he deposed President Ayub Khan and instituted martial law. Now—eight months later—elections were set for December 7, 1970, to choose a National Assembly which was to write a new constitution. General Yahya, who repeatedly expressed his desire to revert to a soldier's life, declared somewhat paradoxically that he would tolerate no resistance to the restoration of civilian rule, even if it took the army to do it. For their part, the members of the leading East Pakistan political party, the Awami League, headed by Sheikh Mujibur Rahman, were in no mood for the continuation of Punjabi rule in any guise. They made no secret of their intention to press for a formal regional autonomy, generally taken as the prelude to secession.

East Pakistan had been racked by demonstrations, street fighting, and violence for months, and any kind of a national calamity on top of all unrest could only widen the east-west gulf.

With the December elections less than a month off, a calamity of the first magnitude began to brew up. At about 9:00 A.M. on November 10, so the meteorological post-mortem goes, a prevailing wind blowing into the southeastern part of the Bay of Bengal encountered a deep depression, or layer of moist air perhaps a mile thick, resting on the surface of the water. The wind and the moisture-laden air, in themselves innocuous, provide the two basic ingredients for a cyclone. Together they make up what is known as an intertropical convergence, whose main characteristic is that the wind imparts a mild surface current to the surface air, sending it to higher levels. The vertical circulation becomes increasingly well defined as the water vapor in the ascending air condenses aloft; in so doing it releases large amounts of heat energy which, in turn, fuels the winds and perpetuates the rotation of the air-water mass so that it feeds upon itself.

High-altitude winds carry off the excess air expelled through the top of the chimney which is created by this action, making room for more moist air to be sucked up. The system becomes, in effect, a huge pump which generates the winds whose whirling action gives cyclones their name.

The budding storm moved from its original position five hundred miles east of Bombay to about six hundred miles southwest of Chittagong—still well out to sea. It moved again—northwest and then north—accelerating in rotational speed into a severe cyclone with a core of hurricane intensity. Early in the morning of November 12 it struck East Pakistan over the Mouths of the Ganges, at a point near their western end.

By this time the winds had reached a speed of close to 150 miles per hour. Traversing the coast line in an east by northeasterly direction, they extended out from the storm center 40 miles north and south. The right half of the cyclone, its most dangerous part, found the offshore islands and chars, sandpapering them with a counterclockwise motion like a giant whirling disk. Rolling up everything in its path the entire distance to the Ganges-Brahmaputra estuary, the storm system crossed over to the Chittagong shore and disappeared north of the city. Like all tropical cyclones, it was accompanied by a heavy rainfall.

It was the salt water, rather than the wind, that killed. Luckily for the western portion of the coast, the sea surges came at low tide, raising the water level considerably, but still below that of the normal high-water mark. Unfortunately for the eastern end of the 80-mile-wide storm track, in the great estuary where some of the most populous islands are located the advent of the cyclone coincided closely with the highest spring tides of 9.9 feet.

One of these islands, Manpura, although only a hundred miles southeast of East Pakistan's capital city of Dacca, never had any means of communicating with the outside world except by boat. There was no warning of approaching tidal bores. A survivor told of hearing a great roar from the southeast, growing louder, near midnight on November 13. He looked out into the pitch-blackness and saw a glow in the distance which came nearer and grew bigger. He realized that it was the crest of a huge wave.

This villager was fortunate enough to have a solidly built house. He and his family were upstairs when the wave broke, coming as high as the second story. His was one of four houses out of 4,500 left standing. Of the 26,000 people on Manpura Island, three quarters were swept away. At no point more than 20 feet above sea level, the island was completely inundated.

Manpura was typical of the tragedy that repeated itself along more than a hundred linear miles of coast line. The wall of water was estimated to be 25 feet high, and there was nothing of any substance to block it on the low-lying islands and shores. Humans and animals were swept away. Houses suffered the same fate or, if their roofs were of corrugated metal, collapsed on their occupants.

With an area of 13,000 square miles and a population of more than a million, Bhola is one of the largest islands in the Bay of Bengal. Its southern half was completely devastated. One twenty-five-year-old man told of losing everyone but his father out of an immediate family of fifteen. All were sleeping when the wave struck. Not one of twenty-five children among his relatives survived; in fact, the greatest toll everywhere was of children, who were not able to cling to trees as were the few adults who saved themselves in this manner.

Hatia Island, near Bhola, initially reported 5,000 dead but believed that the number was closer to 60,000 of its 120,000 inhabitants. Half of Hatia's 242 square miles had been under 20 feet of water for eight hours before the first message was received.

The enormous force of wind and water could be seen on a tiny islet between Hatia and Bhola, where a 500-ton cargo ship had been driven 50 yards inland on her keel. No living person was in the vicinity.

Scenes from the air the next day were indescribable. It was not only that entire villages had disappeared but that there was no stretch of 200 yards without bodies and animal carcasses. They lay everywhere—on sand bars, in the waterways, and festooning the trees. All cattle, most of the buffalo, sheep, goats, and fowl were gone; fishing boats and nets were destroyed—leaving the survivors destitute.

The Pakistan Relief Commission quickly estimated the loss of 225,000 homes, with 100,000 more damaged. A total of 1 million acres lay waste, and at least 800,000 tons of rice were ruined.

As for the deaths, the usual numbers game began immediately, with the government hoping against hope that the count would not continue to rise astronomically. It did, nevertheless: 55,000; 168,000; 200,000; eventually 300,000 became the official figure. The true extent, however, was widely believed to be much greater—possibly 500,000 or even 1 million. If the latter seems unrealistic, it had as its basis

1. Looking toward Costa Rica's Mount Arenal from a hill on which stands a monument to the eighty people killed in the volcano's explosive eruption of July 1968. Eight months later, a 100-foot-thick flow of aa (lava heavily interlaced with solid materials) can be seen advancing westward at no more than five to six feet per day. By late 1969 the flow had completely surrounded this spot. *Photo courtesy Sr. Rodrigo Saenz R., San José, Costa Rica.*

2. The head of the Mount Arenal flow built up a surface 60–100 feet high on a wide front, as shown here before individual crosses to four people who died in the initial eruption. *Photo: Sr. Rodrigo Saenz R., San José, Costa Rica.*

3. In a time exposure, Nicaragua's Cerro Negro creates a celestial display caused by static discharges as particles rubbed against one another and violent temperature changes occurred within the vast, 20,000-foot-high cloud during its 1971 eruption. The outburst was one of the most violent in modern times, since there were no fissures or other vents on the flanks of the cone, and almost all of the energy was concentrated in the main crater. *Photo: Ingeniero Franco Penalba, Estudios Penalba, Nicaragua.*

4. St. Vincent's Soufrière boils and bubbles on December 17, 1971, exactly two months after the eruption of 1971-72 was first detected. The lava islands in the center made their appearance above the surface of the crater lake a month previously. On this same day, scientists of the University of the West Indies (Trinidad) made an expedition by boat to the lava islands to collect samples. View is from WSW. *Photo: Dr. John F. Tomblin, University of the West Indies, St. Augustine, Trinidad.*

5. Looking north over the Soufrière crater lake on March 22, 1972, two days after the eruption ended. The lava island is now well over 300 feet above the surface and 2,000 feet in diameter. In its early stages, the prolonged eruption had given rise to considerable anxiety that it would culminate in a deadly Peléean type of outbreak, as it had in 1902. *Photo: Dr. John F. Tomblin, University of the West Indies, St. Augustine, Trinidad.*

6. The children of Pozzuoli, Italy, may have been delighted when a bathing beach appeared where none had existed before, but their parents did not share their feelings. The volcanic uplift of 1970 displaced a water line which had extended well above the base of the sea wall. *Photo: Robert Citron, Smithsonian Institution Center for Short-Lived Phenomena, Cambridge, Massachusetts.*

8. Everything volcanic short of a major active crater apparently surrounds (and underlies) Pozzuoli. Fumaroles, sink holes, boiling mud baths, hot springs, steaming vents abound, and the area is known in Greek legend as the entryway to hell. Yet the city, in its beautiful location, has been continuously inhabited since antiquity, and the plain in which it is situated is probably the most fruitful in all Italy. *Photo: Robert Citron, Smithsonian Institution Center for Short-Lived Phenomena, Cambridge, Massachusetts.*

7. The Pozzuoli Uplift highlights a mystery: These pillars and porticoes dating from a first-century-A.D. Roman market (also known as the Temple of Serapis) have risen somewhat from their semisubmerged state, together with the rest of Pozzuoli. In so doing, they have raised questions about the fact that they are eaten away by shellfish and marine activity to a height of fifteen feet above the ground line. This points to their having been submerged to a considerable depth sometime during the Christian era. Pozzuoli has apparently had some dramatic ups and downs which have not been chronicled. *Photo: Robert Citron, Smithsonian Institution Center for Short-Lived Phenomena, Cambridge, Massachusetts.*

9. The remains of the Peruvian village of Ranrarhica, adjoining Yungay, buried under a mud and rock avalanche from Mount Huascarán triggered by the earthquake of 1970. The boulders are the natural tombstones of all but 50 of its 1,800 villagers. Ranrarhica had been rebuilt after a similar avalanche in 1962 destroyed every house but one and wiped out 3,500 inhabitants. *Photo: El Comercio, Lima, Peru.*

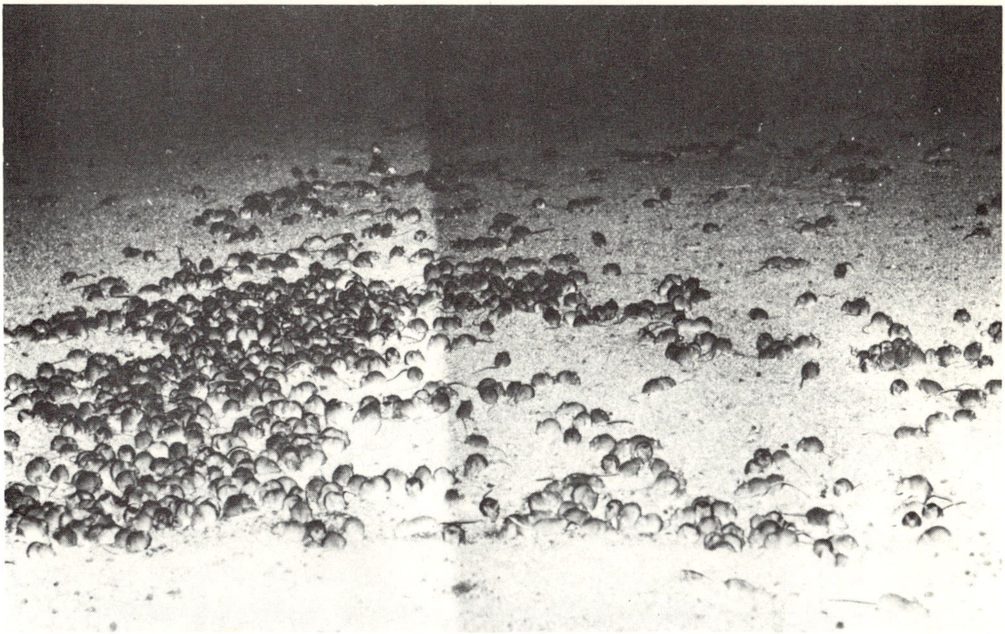

10. Lacking only a pied piper, a plague of mice sweeps across the Mallee region of the state of Victoria, Australia, during 1969. At an estimated average density of 200 rodents per acre, this spot seems to be above average. *Photo: Keith Turnbull, Research Station, Ballarto Road, Frankston, Victoria, Australia.*

11. One of 150 false killer whales (15–20 feet in length, weight about 1,500 pounds) that suddenly and unaccountably "stampeded" out of the water and stranded themselves on the beach at Fort Pierce, Florida, in January 1970. Hauled out into deep water by members of the marine patrol, they dashed back to shore again, until more than 100 perished on the sand. Theories of their behavior, never truly explained, ranged from a sudden drop in sea temperature that set off a search for warmer waters to mass hysteria or some sort of biological disorientation of the "sonar" system by which whales sense depth. *Photo: J. W. LaTourrette, Wometco, Miami Seaquarium.*

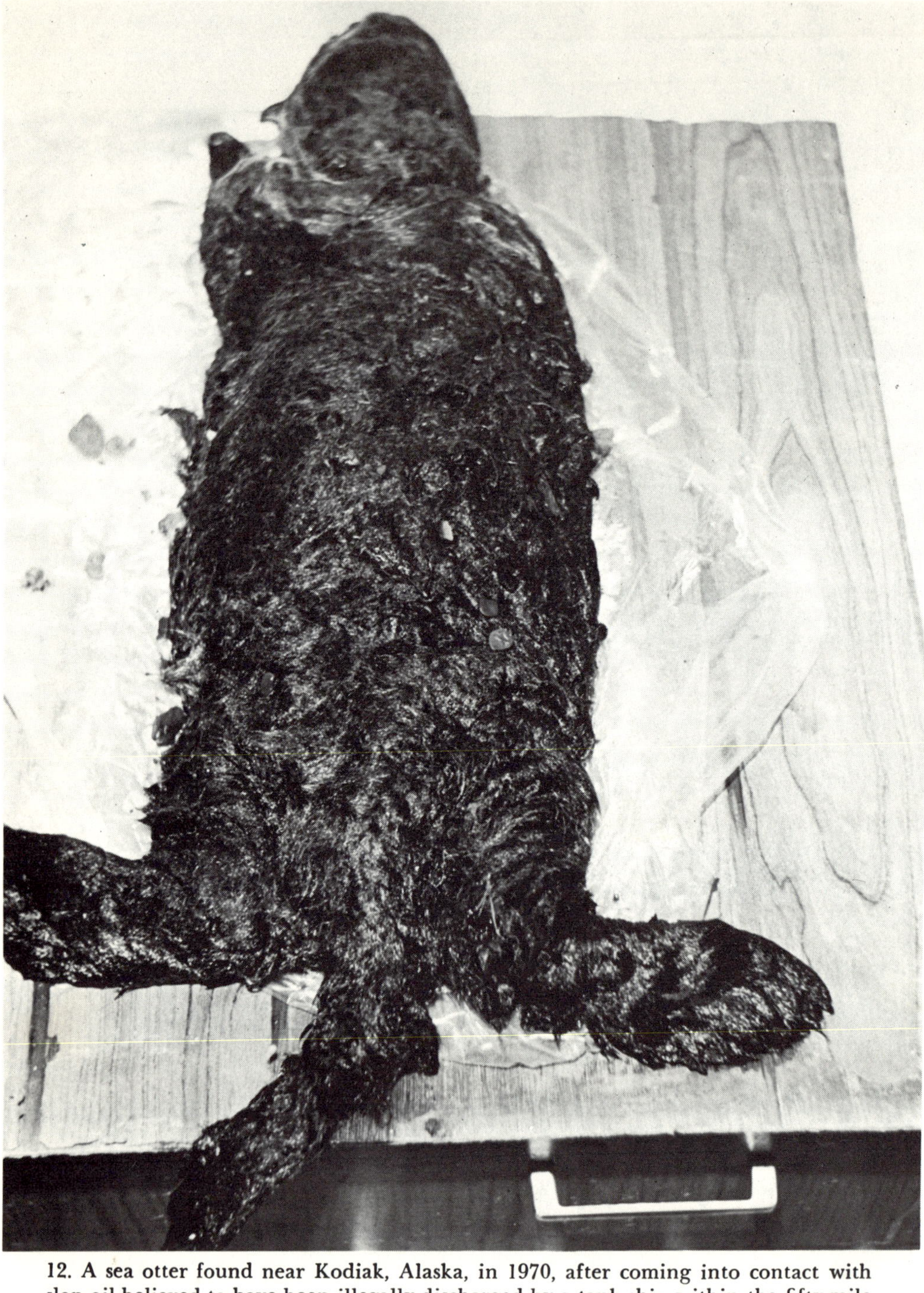

12. A sea otter found near Kodiak, Alaska, in 1970, after coming into contact with slop oil believed to have been illegally discharged by a tank ship within the fifty-mile limit. An estimated 1,000 miles of coastline was affected, and at least 10,000 (and perhaps as many as 100,000) sea birds may have perished in addition to the sea mammals which encountered the oil. *Photo: Environmental Protection Agency (EPA) Official Photo.*

the fact that the aftermath of the Bengal storm surges was in many ways worse than the event itself.

Relief efforts by the Pakistani government began in an unpromising manner and went rapidly downhill. General Yahya, after a two-hour flyover of the disaster area, headed back to Islamabad, the new capital built in West Pakistan. The monumental task remained of getting food, drinking water, medicines, clothing, and temporary shelter to a horde of survivors who had lost virtually everything. They were in imminent danger of dying from starvation, epidemic, exposure, or a combination of all these handmaidens of disaster.

The greatest need was for transportation, which meant helicopters. There were aircraft of this type in West Pakistan, but the government would not release them. The reason quickly became apparent: out of pride, the Pakistani regime would not request permission of its old enemy, India, to overfly their territory. Pakistan would not even relax its attitude when India volunteered military planes, helicopters, and riverboats, as well as two mobile hospitals and medicines. When, some time later, the offer was accepted, Indian planes were refused permission to fly into East Pakistan, and the equipment had to be brought in by truck.

Another dispute broke out over the helicopters made available by the British and Americans. Pakistan insisted that they be flown by its own pilots but backed down when the two Western countries refused to hand the aircraft over unless they were manned by their original crews.

These officious attempts at maintaining a sovereignty that no one was trying to breach were not all that hampered relief operations. The effort was chaotic, without central direction, and what little control there was reflected an unbending bureaucracy. The Pakistani government failed to make known the kind of information which would assist outside relief organizations in making intelligent decisions with respect to aid. Pakistan produces more than enough cholera vaccine for its own needs, and had plenty on hand. Yet the various countries sending supplies were not told this; the first plane from Communist China carried 500,000 superfluous doses of the vaccine, and other shipments by well-meaning nations glutted East Pakistan and took up cargo space that could have been put to better use.

A week after the disaster there was still no effective way to reach the survivors with blankets, tents, high-protein foods, and medical supplies. Virtually the entire water supply in the south had been polluted by saline waves and decomposing bodies, but water purification equipment waited at air terminals with no means of transporting it further. Among the shipments were flat-bottomed boats for negotiating the network of shallow waterways, but no effective plan for utilizing them was put forward. Relief missions, two fully equipped medical teams from Norway among them, sat in Dacca while Pakistani officials debated over where to send them and whether to permit them to operate their own radios.

The United States helicopters, six in all, began to fly out of Dacca, bringing food into the disaster zone. Pakistanis almost mobbed the United States ambassador, who accompanied one of the flights, snatching a bag of rice out of his hands and forcing him to retreat into the rear of the plane.

From Singapore, the British government sent eight helicopters and a task force of 650 troops on two ships carrying World War II landing craft. The small armada anchored in the Bay of Bengal to serve the western part of the region, while the Americans concentrated on the east.

Compared with the urgency of the foreign relief effort, the part played by the central Pakistani government was lethargic. Ten days after the disaster not a single military plane or helicopter from West Pakistan had been flown into East Pakistan. The lone military transport plane stationed at Dacca had not been assigned to relief work for the first seven days, and the Agriculture Ministry's three crop-dusting planes were held back even longer. While the arrival of foreign relief supplies jammed Dacca's airport, the government made no move to curtail commercial passenger flights.

General Yahya Khan was assailed in strong language by eleven Bengali leaders, martial law or no. Their statement charged the Pakistani government with ". . . gross neglect, callous inattention and utter indifference." A telegram sent to the President spoke of "sinister attempts to play down news coverage." "Not one single minister is here," the telegram read in part. "You yourself left the province after a cursory glance at the first flush of news of the tragedy."

Stung by the criticism, Yahya Khan flew back to Dacca on a ten-day take-charge visit. He made a two-day personal tour of the disaster zone during which he spoke with the homeless ("How are you, baba?"), handed out checks, and distributed blankets to which were pinned slips of paper reading, "Gift from the President of Pakistan." From 1,000 miles away, the central government announced the allocation of $105 million more to finance relief operations ($11.4 million had been granted previously).

Meanwhile, the living buried the dead as best they could, although thousands of bodies floated out to sea and thousands more fouled the land and streams. The shortage of water and food was acute in spite of the best efforts of the relief workers and the semblance of organization that emerged out of the earlier chaos. More than a million were believed to be homeless, and pilots reported that some were already beginning to make a move toward domesticity by hanging out clothes to dry.

But for the most part there was nothing left with which to make a start, and the lives of the survivors were again in peril, this time from a slower death. On every mud road and lane people begged and cried out for help, holding cloths to their faces as they walked to keep out the odor of decaying bodies. On the accessible parts of Bhola Island, survivors were frenzied with hunger. Before the arrival of the first foreign helicopters, little could be distributed, and that only sporadically. The arrival of a pickup truck with tins of biscuits was greeted by screaming peasants. After a few tins were given out, the frightened driver started to pull away, people hanging onto the sides of the vehicle as long as they could. One man who managed to get a tin was swarmed over and knocked down.

Some areas were so cut off that nothing was known of conditions for a week afterward. More than half the offshore islands were written off by Pakistani officials as hopeless cases.

In Dacca the Bengali leaders continued preparations for the December 7 elections, warning Yahya Khan not to use the cyclone as an excuse for postponement. They maintained bluntly that thousands more lives could have been saved if the government had reacted promptly and effectively within twenty-four hours. Their assertion was supported by admissions from the governor of East Pakistan that, after almost two weeks, there was still no government program to co-ordinate relief activities—either those of their own agencies or of the many foreign groups which were performing most of the rescue work.

Gradually the waters receded and the islands and mainland shores re-emerged. Ironically, the cyclone had not changed any of them, as do the gentler monsoon floods and tides. The area was exactly the same in topography and outline; only the people and the animals had disappeared, and with them their habitations and cultivated fields.

Whether 300,000 or 1 million perished—and the true figure probably lies somewhere between—the grim mood of the Bengalis was deepened by the knowledge that the human disaster might have been not only alleviated but perhaps prevented. It seems strange that this impoverished people, with no resources but their own hands, would think it possible to protect their vulnerable lands from such irresistible forces as wind and water. But they had been told that it was indeed possible, and on excellent authority.

In 1960, after the cyclones of that year had taken an undetermined number of lives, American specialists had been invited to East Pakistan to make a study of the situation. The former head of the National Hurricane Center in Miami spent four months there and made a number of recommendations. Among them was one for an efficient warning system in an area where it was not unusual for an island with 250,000 inhabitants to have no electricity and only one battery-powered radio. Another was that earth platforms be built on the islands as simple but effective places of refuge from tidal bores, and as sites for schools and public buildings.

Nothing came of these recommendations. In the next ten years one radar station was built to watch storms as they moved up the bay, but no way of spreading the alarm to the populace was devised. Neither was anything done to provide the sanctuary platforms. As a result, the cyclone of 1970 was a repetition on a larger scale of past disasters.

As they counted their dead, the bitterness of the Bengalis was intensified by the awareness that a simple works project utilizing plentiful local labor and earth fill had been neglected in favor of an extravagance such as the new capital of Islamabad. Rising from scratch on a 350-square-mile site in the northern part of West Pakistan, this city of 70,000 had been constructed at a cost of $300 million between 1960 and 1970, with plans to expand its population by another 200,000 over the next decade. The contrast between this expensive monument and their own life-and-death needs was not lost on the Bengalis, and the fact that a new capital was also built in Dacca to serve as the seat of the National Assembly for half of each year did little to redress the balance.

Nothing that the central government of Pakistan did following the disaster was regarded as more than belated, feeble, insincere, and, at best, politically motivated. In Dacca, not far beyond the radius of destruction, the Bengalis pursued their single-minded drive toward self-determination. Rallies, parades, electioneering did not stop even during the official national day of mourning. The December 7 elections were held as scheduled, except for several districts which were so desolated that nothing approaching a vote could be organized.

The Awami League captured the largest bloc of seats, and an absolute majority, in the National Assembly, much more than the Punjabis had bargained for. Twice thereafter, Yahya Khan postponed the first meeting of the Assembly and, therefore, the writing of the constitution which was its task. General strikes, protests, and street violence broke out. Negotiations between Sheikh Mujibur and General Yahya broke down, and the sheikh declared the independence of East Pakistan.

The invasion of East Pakistan by 70,000 West Pakistani troops, the bloodshed, the flight of 7 million Bengalis as refugees into India, and the eventual intervention of India itself were the principal events in the creation of the state of Bangladesh. The country went through another agonizing experience, this one prolonged and general, before rebirth as a new nation in 1972.

The problems facing Bangladesh are horrendous, especially after the destructive internal fighting, and there are few illusions about a future which has automatically become rosy with independence. Substantial foreign aid is programmed, with the United States having contributed a total of $215 million (partially through the UN) as of mid-1972. Other practical measures were also taken, such as the chartering for six months by the U. S. Government of the 115,000-ton supertanker *Manhattan,* largest ship in the American merchant fleet, to serve as a floating grain silo anchored off Chittagong. This was necessary to help by-pass damaged and congested port facilities and speed the distribution of food into the country's interior. Transportation was so seriously disrupted by the war that heroic efforts were called for to stave off malnutrition and possibly famine. Among the facilities totally destroyed were 276 bridges.

While plans and improvisations based on modern technology got under way, the Bengali farmers of the southern flatlands took destiny into their own hands and began to change the map. The earth platforms at last made their appearance on the chars and offshore islands. The work started under the direction of the International Red Cross, whose teams had been continuously in the area since the 1970 disaster.

With the support of an initial $1.5 million allotment from the Bangladesh Ministry of Rural Development, the League of Red Cross Societies supervised the construction of 235 *killas* (Bengali for shelters) in strategic locations. Gangs of up to 1,000 village laborers, recruited at a daily wage of forty-two cents or the equivalent in food, complete a *killa* in about two weeks, moving the earth in baskets carried on their heads. Each platform measures about 100 by 150 feet on top, enough space for more than 1,000 persons and their domestic animals, raised 20 feet above ground level.

The slopes of the earth structures are planted with grass and with banana and palm trees to prevent erosion. A brick building is planned for each to protect the people from the elements.

The *killas* will have provisions for six days, the usual length of time it takes to get relief to an area by boat. In the future, each *killa* will also be the center of a development complex which will include a school and co-operative store.

Rounding out the recommendations made long ago, and just as long ignored, warning and rescue teams equipped with radios will operate in conjunction with these centers of refuge.

The *killas* and the activities associated with them should bring the traditional village life of southern Bangladesh to a focus which it has never had before. Common peril has finally given rise to common action. Bengalis who periodically met disaster as individuals, and just as often succumbed to it in the same way, are now combining their strengths against one of the forces which once defeated them. It was not apathy, ignorance, or weakness that put off the day for so many centuries, but a lack of national identity and sense of mutual interest. The change, when it took place, was not in the physical world of Bangladesh but in the minds of its people.

A Few New Wrinkles in the Earth's Skin

(Landslips, Rockfalls, Mud Slides, and a Rootless Island)

Earth is as capable of minor annoyances as it is of awesome rampages. Even its petty phenomena, however, reflect the great catastrophes from which they differ only in degree. Some of these practical jokes of nature are harmless, others border on the perilous, and still others cross the thin line into destruction of human works and sometimes of life as well. A hailstorm in June is not unusual; one in which the hailstones range from almost an inch to two inches in diameter is not only unusual but serious. Milan was subjected to such a bombardment for a total of twenty-two minutes one June afternoon in 1971. The city came through with no consequences to speak of, but the agricultural fields of the Lombardy plain experienced severe damage, and floods were reported in certain areas. There is no lack of such mischief, from below as well as above ground, and information on incidents of this relatively small magnitude is communicated continually from almost everywhere in the world, as these few examples will show.

MAGHAR, ISRAEL, January–February 1969 . . . An intensive landslide was recorded in this village on the steep southern slope of Mount Hazor in Galilee. This side of the mountain is extremely faulted and consists of Lower Cretaceous soft marly strata.

The landslide activity lasted on and off for four to five days. The movement was shallow, interrupting only the soil cover and upper marly sequence.

This is the third time that the village has been damaged by a landslide. The other occasions were in 1894 and again in 1928.

URKUT, HUNGARY, March 19, 1969 . . . Between the fourteenth and nineteenth of the month, a landslide presumably due to heavy rains occurred on Mount Bocshoc here. The landslide destroyed 80 meters of highway between the towns of Ajka and Urkut.

NORSÄLV, SWEDEN, April 12, 1969 . . . A clay and sand landslide covered an area 300 by 400 meters, cutting off a road, blocking the Norsälv River, and causing waves 4 to 5 feet high.

SALO, FINLAND, May 19, 1969 . . . Two thousand cubic meters of a 5-meter-high escarpment of the Salo River slid for a distance of 50 meters within the city limits here. A year or two ago the land on the opposite side of the river sank 10 meters.

DAVAO, REPUBLIC OF THE PHILIPPINES, May 1969 . . . An earth mound discovered early this month near the northern bank of the Mahanob River, about five miles west of the town of Baganga, rose rapidly to a height of 5 meters in a little over three weeks. At about the same time earthquakes were felt in the region, and sounds similar to the crushing and sliding of rocks were heard.

The mound is about 50 meters long and ends abruptly in a steep slope where it meets the plain. It stands on the base of one of the ridges that trends eastward, perpendicular to the main mountain range. A rivulet which had previously flowed at the base, hugging the foot of the ridge, has been uplifted to expose the river bed.

On top of the mound are several cracks parallel to its long axis. The trees with which it was originally crowned are still standing but inclined more or less in an eastward direction.

The absence of any odors, such as sulfur, of a temperature increase in the ground, or of vegetation discounts the possibility of a volcanic origin for the phenomenon. The land rise is attributed to slumping, a kind of slope failure, precipitated by the washing away of the foot of the slope by the stream, overloading of the ridge by persistent rains, and, finally, earthquakes.

It is believed that the same factors acting on other ridges may result in similar slumping until a new stability is attained.

CASTROVIRREYNA, PERU, June 1969 . . . An entire district here, in the Peruvian Department of Huancavelica, is slowly sinking, owing to a geological phenomenon that has caused cracks in the earth along a circle with a 3-mile circumference. The adobe houses are slowly inclining and may collapse at any moment, burying their occupants.

CORUSQUEIRA, THE AZORES, July 27, 1969 . . . Several tons of rock and sand slid down a 60-meter-high escarpment, destroying several homes in its path before rolling into the sea. This took place on June 25, followed on July 27 by a repeat performance which caused no damage.

TATABANYA, HUNGARY, August 15, 1969 . . . Heavy rains caused a mudflow near here which covered the M-1 Autostrada to Budapest to a depth of 30 centimeters over 300 meters of its length. Another mudflow covered the road near Vertesszolos at the same time.

HUEHUETENANGO, GUATEMALA, August 28, 1969 . . . A landslide consisting of mud and stones and caused by heavy rains severely damaged the villages of San Antonio and Las Nubes. Eighty people were believed killed.

IMESSI-ILE, NIGERIA, September 1, 1969 . . . During heavy rains this afternoon a 200- to 300-ton block of granite, 5 meters high and 8 meters in diameter, slid from the side of an inselberg here in the western part of the country. Inselbergs are isolated mountains partially covered by debris from their own slopes.

The boulder cut a path 80 meters long and 8 meters wide. The upper portion of the gradient down which it slid was at an angle of 30 degrees, and the lower portion 10 degrees. The sound of its passage was heard up to 5 kilometers away and caused widespread alarm, as the "mountain" was thought to be angry.

Shearing off a tree, pushing vegetation and soil before it, the boulder was believed to have begun its slide after an increase in its effective mass through saturation of attached soil and greenery.

MEIJE, FRANCE, September 2, 1969 . . . A rockfall from the southwest side of the 3,983-meter-high Meije peak, fifty kilometers southeast of Grenoble, almost completely covered the three-kilometer distance to Etançons glacier toward the south. The height of the fall was about 1,400 meters, and the fallen rockpane seemed more than 300 meters high and 100 broad.

JUNEAU, ALASKA, August 15, 1970 . . . Lake Linda, a super-glacial body of water, 4,000 feet above sea level on the Juneau ice field, began to drain two days ago and had completely emptied itself by today. It was the first time that an event of this kind had been witnessed by a trained observer, a research affiliate of the Glaciological and Arctic Sciences Institute of Michigan State University.

It had been thought previously that this type of lake disappeared in a flash, as in Peru, where superglacial lakes that drain either because of an avalanche or earthquake (see "The Ice Falls of Huascarán") cause catastrophic flooding and enormous loss to life and property.

This was not true of Lake Linda which, 125 meters long and 50 meters wide at the water line, and 10 meters in depth at its deep end, began to drain slowly. It lay in a basin with 70-degree walls at three sides, 90 per cent covered with snowbergs, and with the long axis of its oval shape parallel to the prevailing southeast wind.

There was no apparent reason for the draining. A nearby falls stopped running just prior to and during the drainage. But since it was about 10 feet higher than the water surface of Lake Linda, it was obviously not an overflow drain for the lake.

From about 5:00 A.M. on August 13, the water began to drop at the rate of 3 to 4 inches per hour. There was heavy wind and rain, the latter compensating somewhat for the drainage but not sufficiently to check it. The bergs in the lake began to settle and crack, a strong organic stench of undisclosed origin arose, and by 11:00 A.M. on August 15 the lake bed was completely empty of water. Everywhere on the bottom, icebergs had come to rest, cracking into a pattern of concentric ovals and altering the appearance of Lake Linda beyond recognition.

DUNAFÖLDVÁR, HUNGARY, September 15, 1970 . . . A great landslide occurred today along a semicircular arc estimated at 700 meters long by 120 meters wide in this small town near the Danube, some seventy-six kilometers south of Budapest. The slide took place near the railroad bridge, compressing a great deal of the high bank of the river into the water. A vineyard, a fruit garden, two small dwelling houses, and a country house were moved downward 20 to 30 meters, but only part of one of the houses collapsed. There were some slight injuries but no deaths.

The soil is now deep in the river, some 50 to 60 meters from the bank and about 300 meters in length. It has not stopped shipping at this point on the Danube, since the width is approximately 700 meters.

According to a rough estimate, more than 1 million cubic meters of material were moved by the slide, making an intense rumble in the process. The compressing force of the landslide created two islands in the river channel, roughly parallel to one another for about 300 meters. They are 3 to 5 meters in height and of widths which vary between 4 and 12 meters. They bear deep fissures, are composed in part of tilted blocks of clay, and display many characteristic Danubian shells and snails on their surfaces. It is obvious that these "miniature mountain chains" are of the same composition as the river bed and were pushed up from it. The entire scene, particularly because of the soil deformation, resembles an area after a violent earthquake.

Immediately after the landslide an enormous cloud of dust rose into the air. Upon the sudden emergence of the islands, some strong waves originated in the river waters, much like small tsunamis. Despite the width of the river, the waves were strong enough on the opposite bank to cause panic among the residents there.

A preliminary estimate of the damage showed it to be the equivalent of at least 1 million Hungarian forints. Subsequent findings were that five houses were partially or totally destroyed and a wine vault collapsed. The hutches of a nearby rabbit farm, however, were preserved, and a dwelling place survived between two deep chasms that opened up.

The causes of the landslide were the abundant rains of the summer and the fact that the area behind the earth which gave way is a catchment basin. Water loosened the layers of soil under the surface, beginning the process.

WOLENCHITI, ETHIOPIA, August, 1970 . . . The ground here, some eighty kilometers east southeast of Addis Ababa, has again fractured in a characteristic pattern of deep fissures. Why they appear is not known scientifically, although the location is in the main Ethiopian rift and in the midst of an extensive belt of deep earth faults.

The rainy season aggravates the ground fracturing, although in the dry season of 1969 a farmer lost his ox and plow while tilling the soil over a hidden fissure.

Members of the staff of the Smithsonian Astrophysical Observatory and of the faculty of Haile Selassie I University have visited the area and are engaged in an investigation of the causes of the recurrent fracturing of the Ethiopian rift floor.

CHUNGAR, PERU, March 18, 1971 . . . An avalanche of rocks loosened by weathering and rainfall crashed from a mountainside into

Lake Yanahuani, causing a wave estimated at 60 feet high that completely destroyed the copper- and lead-mining community of Chungar, located in the Department of Pasco, ninety miles northward of Lima.

Eyewitnesses say the mass moved at speeds close to 200 miles an hour. The wave that destroyed the town killed 200 people.

WINDWARD PASSAGE, July 4, 1969 . . . An island floating in the Caribbean with no apparent destination was sighted today by the United States destroyer escort *John D. Pearce,* sixty miles south of Guantánamo Naval Base. The island was about 45 feet in diameter and was topped by ten or fifteen trees, 30 to 40 feet tall.

(A week later the island was again sighted, this time by the Coast Guard cutter *Hamilton.* In a more detailed description, the trees were said to be like palms but with no tops. There did not appear to be any life or foliation; the island was brown in color and looked to be held together by a mangrove type of matting. At ten miles' distance there might have been some earth on it, although the ground level resembled dead foliage or grasslike material. The trees were embedded in this surface. No roots were visible.

(As a team of Smithsonian Institution ecologists prepared to visit the floating island by helicopter, a search in the area where it was last seen failed to turn up any sign, and the expedition was called off.)

"Whatever Wasn't Falling Down Was Burning Down"

(The Managua Earthquake: Nicaragua, 1972)

Three times within the last century Nicaragua's capital city of Managua has paid the price for being in the wrong place. It stands squarely over a geological fault that slipped in 1885, in 1931, and again in 1972 to cause devastating earthquakes. After the first two disasters, the survivors buried their dead, rebuilt their homes, and resumed business as usual. After the third, they seem prepared to follow the same path.

There is no compelling reason why Managua must be made to rise repeatedly from its ashes only to be leveled once more. If it had not been for the bitter rivalry in the nineteenth century between León and Granada, then Nicaragua's two leading cities, Managua might well have remained a minor town on the south shore of the great salt-water lake from which it takes its name. It was founded by the conquistadors on the site of an Indian village and could have continued in obscurity, or been abandoned in time, if it had not been chosen as the compromise capital of Nicaragua in 1855.

The danger of earthquakes in Nicaragua was known to be so great that, with the memory of the 1885 disaster still fresh, the United States decided in 1901 to build the canal connecting the Atlantic and Pacific Oceans in Panama instead. Thirty years later Managua again suffered grievous damage, this time with the loss of 2,000 lives, in a temblor that lasted only six seconds. There was still another earthquake in 1968, but it measured a mere 4.5 on the Richter scale, a magnitude that usually results in a very local and limited effect and was inconsequential by Managua standards.

It was a different story on December 23, 1972. Just thirty minutes past midnight, local time (063136 Greenwich Mean Time), the city of 325,000 people was struck by an earth shock that registered 6.25 on the Richter scale and originated at an extremely shallow depth. The energy released was roughly equivalent to that of an atomic bomb fifty times as powerful as the one dropped on Hiroshima. Approximately forty-five minutes later, there were two aftershocks within ninety seconds of one another, registering 5.0 and 5.25 respectively.

Managua collapsed. Fires broke out and quickly got out of control. "Whatever was not falling down was burning down," said one observer.

Thousands died, and additional thousands were injured. Three quarters of the buildings were destroyed and another 15 to 20 per cent were so severely damaged as to be rendered unsafe or uninhabitable.

At least 100,000 survivors fled into the countryside. On Christmas Eve, total evacuation was ordered. Ruptured water mains and sewers were inviting typhus. Electricity and phone service were out, leaving auxiliary-powered short wave radio the sole form of communication.

There were scores of aftershocks, and one strong tremor toward noon, but although they continued for several days, there was not much they could add to the almost total damage that had already been done. The once sparkling lakefront city lay shrouded in the dust still rising from its ruins and the smoke of hundreds of fires in the rubble. Scores had died at the El Retiro Hospital when it collapsed. The Presidential Palace was seriously damaged, showing great fissures. Below on the same hill, the pyramidal Intercontinental Hotel lay dark and empty. An entire floor had been occupied by a reclusive American billionaire who, only a short time before, had been besieged by newsmen anxious to report his every move. He had disappeared within hours after the earthquake, and it was not learned until some days later that, driven by a lifelong fear of natural disaster, he had been flown off to London in his private jet. He had taken off from the airport where seriously injured Nicaraguans lay awaiting transport to hospitals in other countries. The main hospital in Managua had been declared unusable; others were ill equipped to handle medical emergencies, even in normal times.

While the city flamed and smoldered, many who had lost their homes sat silently in the middle of the streets. Others lay on cots and mattresses in the roadways beneath a bright moon and starry skies. Rats, dispossessed by the shattering of the buildings, scampered by. The injured tried to bind their wounds with dirty bandages. In the poor Oriente district, the crushed and broken houses had the haphazard look of overturned sugar cubes. For the most part it was quiet, except for the occasional barking of a dog or the sound of a portable radio telling Nicaraguans to keep calm.

On Christmas Day the Nicaraguan government cut off food supplies to force survivors to leave the city before decaying bodies under the rubble caused an epidemic. The death toll was estimated at 3,000 to 5,000, figures which were revised a few days later to 4,000 to 7,000, with 15,000 to 20,000 injured. The homeless numbered 200,000.

The United States and Nicaragua's Latin American neighbors began rushing aid and supplies into the country. The World Health Organization and twenty-four nations, including many private and religious relief agencies, took part in the over-all effort. The earthquake had compounded an already disastrous food shortage in Nicaragua, brought about by a drought which had destroyed about 80 per cent of the corn crop, 35 per cent of the bean and rice crops, and 20 per cent of the cotton. The U. S. Agency for International Development had been planning to send food relief when, suddenly on December 23, the need became immediate. Forty-five thousand tons of foodstuffs were earmarked, but since they were to come aboard ship, their arrival would be delayed until mid-January.

The interim measures, however, were swift and effective to a point. The United States sent an average of twenty-seven transport planes daily, with food, medical supplies, communications equipment, water purification plants, and demolition gear. Two American field hospitals were set up, treating almost 1,600 patients between them in a few days. The WHO sent special epidemic-combat teams of physicians and nurses to prevent outbreaks of cholera and typhus. Other countries, France among them, established field hospitals. Huge stockpiles of food and other life-giving supplies were accumulated in central locations guarded by Nicaraguan troops against looters.

It was here that the rescue program broke down, as it often does in similar situations. Distribution was lacking. In Masaya, a town thirteen miles to the northwest which had become the principal refugee center, Red Cross volunteers were confronted with the task of apportioning such typical shipments as 50 pounds of rice, 100 pounds of beans, and a few bananas among 300 ration-card holders. There were the usual rumors of corruption and diversion to the favored few. They led almost directly to the death of Roberto Clemente, the star outfielder for the Pittsburgh Pirates, and the most popular sports figure in the history of his native Puerto Rico. He was the leader of

Puerto Rican efforts to aid the Nicaraguan victims and insisted on accompanying a planeload of relief supplies to keep them from falling into the hands of profiteers. He was lost when the plane crashed on take-off from San Juan, Puerto Rico, on New Year's Eve.

A few key Nicaraguans publicly expressed the opinion that the donors of the relief supplies should undertake direct distribution. Some of the national and private agencies began to do so. One reaction by the United States was to assign four CH-47 helicopters from Fort Hood, Texas, to airlift supplies to outlying areas where refugees were being sheltered in homes and public buildings. Each of the CH-47s has a cargo capacity of 10,000 pounds and a range of 300 miles.

Four days after the earthquake, Managua was described as a scene of putrid devastation. A semblance of order had been restored, and the evacuation continued. People still dug in the rubble. The bodies of more than 2,000 victims had been buried in mass graves outside the city, and other bodies were burned where found.

In the square mile of downtown area that had been virtually destroyed, U. S. Army engineers and Nicaraguan rescue teams were at work clearing away the rubble. The clearance of ruins was expected to take at least eight months, but the use of dynamite and bulldozers at the outset generated friction among Managuans and accusations that the authorities were callously razing everything, without regard to what could be salvaged or restored. They were also supected of keeping fires going in order to encourage the departure of those stubborn residents who refused to leave.

Spurred by these grievances, the art of political dissent revived faster than anything else in Nicaragua. The leader of the country since 1933, General Anastasio Somoza Debayle, had recently completed a five-year term as President. Unable to succeed himself under the constitution, he had governed since the spring of 1972 through a three-man provisional junta, with two of its members drawn from his loyal Liberal Party. The opposition Conservatives had lent themselves to this arrangement by furnishing the third member of the triumvirate and found themselves with even less voice than before. The general, who retained control of the armed forces, ruled uninterruptedly from behind the scenes.

The Conservatives made themselves heard from the side lines, complaining that they had been excluded from the National Emergency Committee organized to conduct relief operations. They also protested against the military dominance of the situation and the failure of international aid to reach its destination. They were particularly incensed that all refugees in need of food had been instructed to display a red flag, which happened to be the symbol of the Liberals.

Defending the role of the military, the government maintained that the opposition would have been ineffectual if the mobilization of resources had been in their hands. The failure to distribute relief supplies widely was attributed to a shortage of transportation which, the authorities said, they were trying to alleviate by renting the necessary vehicles from Guatemala and Costa Rica. In this connection, there were reports that the Conservatives had many trucks available, but the government had apparently not called upon its political opponents for help. As these charges were made on both sides, there was looting for food, and the city echoed with gunfire after dark.

The grim picture was brightened somewhat by the news that 90 per cent of the country's industry remained intact and needed only electrical power to resume production. This fact had more than just economic significance. Most of Nicaragua's manufacturing plants are located outside the center of Managua, on the road leading to the airport, and far enough from the fault zone to have been saved. If Managua itself were relocated a few miles away from its present site, the chances of earthquake damage would be considerably reduced—if not eliminated. Historically, no effects have been felt outside a 25-mile[1] radius from the convulsions that have periodically ruined Managua and killed its citizens.

In the days immediately following the latest catastrophe, some heartfelt doubts were expressed in the highest government quarters that Managua could rationally be rebuilt where it once stood. As the impact receded somewhat, so did the outlines of a tentative program for total removal elsewhere. The next proposal put forward was to survey the line of the geological fault, transform the terrain above into an uninhabited greenbelt, and reconstruct the city around it on firmer ground. General Somoza counseled a waiting period of six months before actual rebuilding started. In the absence of a logical plan, Managuans left to themselves will inevitably display the strong if unwise determination of people everywhere to reshape their disrupted lives in the image of the familiar past and let the future take care of itself. There is precedent in Central America for changing the seat of government after earthquake damage. In 1776, Guatemala aban-

[1] On February 24, 1973, another strong earth tremor was felt in Managua, as well as five miles north and eight miles south of the city.

doned Antigua for Guatemala City for that reason, thus providing itself instantly with a new capital and Nicaragua ultimately with an example.

Less than a month after the Managua earthquake, American geologists published reports that raised hopes on the possible development of a reliable early warning system to pinpoint both the location and probable strength of budding earth tremors. The reports confirmed the research of Soviet scientists, based on fluctuations of subterranean pressure waves (comparable to sound waves) and shear waves (those producing motion at right angles to the direction of travel). When the ratio of the velocity of these two quake-generating waves drops and then returns to normal, an earthquake takes place. The ratio begins its drop three months before a moderate quake and one month before a small one.

At first it was thought that this effect might have been peculiar to the region north of Afghanistan where the Russian tests were conducted. However, the American scientists were able to make similar observations in the Adirondacks, giving rise to the possibility of warnings of "months or longer" for magnitudes greater than 6 on the Richter scale.

The system was described cautiously, since there are many obstacles to be overcome. Among them are the establishment of many seismic detection stations in earthquake-prone areas and a validation of the method for quakes which originate deep in the earth.

Even if the system proved effective, there are reservations concerning its value to a city such as Managua which up to now has ignored earthquake-proof construction. The British magazine, *Nature,* asks in its issue of January 12, 1973, what use earthquake detection would be where buildings are so vulnerable to tremors and ". . . afterwards there is no city to which the temporarily evacuated inhabitants may return." It is a question which other cities, larger than Managua and in countries infinitely more prosperous than Nicaragua, could also ask of themselves.

The 7,000-Year Sleep

(The Eruption of Helgafell: Iceland, 1973)

Until 1963 there were fourteen islands in the Vestmannaeyjar (Westman) group some ten miles off the southwestern coast of Iceland. Then there were fifteen, when an undersea eruption lasting four years built the one-mile-square isle of Surtsey and topped it with a cone 480 feet high.

Like their motherland, the Vestmannaeyjars are all of volcanic origin. The only one with a continuous history of habitation is Heimaey, approximately four square miles in area, which sits in the midst of the ice-free waters that comprise some of Iceland's richest fishing grounds. It also has a volcano of its own, known as Helgafell, a type different from the steep Vesuvian cones and gently sloping cupola, or shield, craters which are also found throughout the rugged country.

Helgafell is a huge fissure in the earth, similar in form to Laki, which figured in the mainland eruption of 1783.[1] When active, fissure—or crater row—volcanoes vomit lava streams out of a series of vents along the line of the rift, or even directly from ground level without the intermediary of craters.

Laki has more than 100 relatively small cones. Helgafell has far fewer and differs in another important respect: it lay dormant for seven thousand years.

The Vestmannaeyjar Islands were first settled by runaway Irish slaves in the early years of the Christian Era. The Scandinavians came later, and by the sixteenth and seventeenth centuries Iceland had fashioned enough of an existence for itself out of the hard land and chill waters to become prey for English, Gascon, and Algerine pirates. In one ironic chapter of its turbulent history, a raid by

[1] Iceland's volcanoes alone are estimated to have produced one third of the entire world's lava flow since the mid-sixteenth century. Laki's emission in 1783 of 218 square miles of lava was enough to form a cube 7.5 miles square on each side.

Algerian corsairs carried off 400 of the islanders into the slavery from which their remote ancestors had escaped.

As Iceland stabilized politically and socially, Heimaey grew and prospered. Its "capital," hardly more than a village on the northern tip, took the name of Vestmannaeyjar. By 1973 the island accounted for 20 per cent of the country's fishing catch and 12 per cent of its exports. A fleet of eighty boats, four freezing plants, and three factories for the processing of fish meal and oil supported a population of 5,500. As one measure of Heimaey's affluence, there were 800 cars on the island, and on Sundays the custom was to drive out along the only motor road, winding five miles along the shore line, and park at the lighthouse for the view.

The Sunday peace of January 21, 1973, was disturbed by a swarm of earthquakes recorded as originating within six miles of Helgafell. They continued for the next two days, more than 200 in all, but none registered above 2.7 on the Richter scale. There was no alarm on Heimaey; earthquakes are frequent in Iceland, especially in the western part of the southern lowlands from which the island is separated by only a few miles of water.

Just before 2:00 A.M. on Tuesday, January 23, the occupants of the house nearest Helgafell, a family owning one of the three farms on Heimaey, were roused when the grass outside their windows began to burn. A rift 1.1 miles long had opened on the east side of Helgafell and was spewing out smoke and flame. Molten lava began flowing eastward down the ridges and into the sea. The sky was alight and could be seen for hundreds of miles.

The farm family fled across superheated ground. Their home was later set on fire and destroyed by the lava.

Hot ashes and lapilli began to fall on Vestmannaeyjar. Awakened by the red glow and continuous rumbling punctuated with explosions, the townspeople turned out into the streets. It quickly became apparent that the eruption was making the island untenable, and the mayor gave the order for evacuation.

Reykjavik and the nearby American base at Keflavik are some ninety miles northwest of Heimaey, and rescue operations were organized with minimum delay. First to be taken off were the old and sick, who were brought out by United States helicopters from Keflavik. The rest of the population was ferried out mainly by fishing craft and by gunboats of the Icelandic Coast Guard which were not far off to keep watch on British and West German trawlers in the perennial dispute over fishing rights in the area. A few planes were also pressed into service, landing and taking off in the intense heat and fallout that enveloped the tiny airfield at Helgafell's foot.

Toward midday of January 24 the evacuation was complete. No casualties were reported, with the exception of a horse which fell into a crevasse that opened beneath it. About 300 people remained, all officials and firemen apart from a few die-hards among the young and very old. From time to time fishermen returned briefly to pick up equipment they had left behind.

As plans were being made to bring out valuable property, including $15 million in fish products stored at dockside, the eruption grew stronger. The fissure extended to almost two miles on a straight NNW-SSE line with lava flowing into the sea at both ends, effectively splitting off the eastern quarter of Heimaey. There were fears that the lava would creep over the brow of the crater row and into Vestmannaeyjar, lying only half a mile to the west in a shallow bowl. The fact that the lava had found channels which diverted it north and east into the sea posed another problem, however. New islands began to form off Heimaey, threatening the pipes in the sea bed which brought water and electricity from the mainland.

The lava fall from the cliffs into the water sent up huge vapor columns along the island's east coast. From the fissure itself, lava fountains leaped as high as 375 feet. On the first day a cloud of smoke, gas, and ash three miles high hung in the clear blue skies over Heimaey. In the next two days it grew to 4 miles in height, covering an area 37 miles long and 12 miles wide at its greatest breadth. By night the fissure was like a blood-red incision in the earth's skin, jetting and gouting flame along its entire length while the town below was backlighted and thrown into bone-white relief against the dark ridge.

While the threat of a sudden end in molten lava hung over it, Vestmannaeyjar was being steadily smothered by the fall of tephra. Three days after the eruption began, roads were choked and nearly impassable. The airfield was covered to the depth of an inch, and the lower portions of the red and green pitched roofs of the houses were blackened. Deeper drifts accumulated in places. The continued lava flow had raised the temperature of the sea waters to 111 degrees Fahrenheit and had destroyed at least five houses on land. All but a few of the 300 rescue workers were ordered off by the government.

As of Friday, January 26, a new concern had arisen. Lava had filled in and closed up the southern end of the fissure, and a new crater had opened in the northern segment. The eruption was concentrated in a 1,200-foot stretch, lying roughly in the middle of the fissure, with half a dozen smaller fountains above it at about 300-foot intervals. It was feared that the closure would create new volcanic pressures which might open up other rifts. A fresh-water well recently drilled near the harbor was said to be throwing up steam. This suggested underground turbulence beneath the center of town.

With the shortening of the fissure, the lava now began to flow mainly northward, forming a bench that projected at least 1,500 feet out from the coast. There was still the possibility that the flow could change direction to engulf the town, and bulldozers had piled up huge earth barriers against this eventuality. But while the flow of lava into the sea spared the town, it created new troubles. In a few more days the harbor was to be nearly closed off.

The main crater is nearly 1,500 feet across. It built itself a cone to the height of 330 feet within a week of the outbreak, and to more than 500 feet as the eruption continued into mid-February. At that point the lava flow had reached more than a square mile in area and about 75 million cubic yards in volume. It also began to undergo a change in composition, from the low-viscosity, fast-flowing hawaiite to the slower-moving, higher-viscosity basalt which is normal for Iceland. Accordingly, the average production of lava was slowed from about 70 cubic yards per second to 40.

Almost a month to the day after the start of the eruption, fresh activity was reported from the fissure. The future of Heimaey is in the balance. Although no lives were lost, the life of the island has been throttled. Most of the refugees have found homes with relatives on the mainland, the fishermen have based their boats in other ports, and the factory workers will probably find work in other fish-processing plants. Nevertheless, the displacement of more than 5,000 people out of a total population of only 200,000, and the dislocation of their productive labors, has been a severe blow to Iceland's economy. Helgafell has been described by the Iceland press as the country's greatest disaster of the century.

Even if the eruption subsides to a relatively low level, every east wind that blows could blanket the island in ash and fumes, making it virtually uninhabitable. If Surtsey is any indication, the Helgafell fissure could remain active for years.

Gauging their chances for a return home, the exiled islanders have already taken one step toward the hoped-for pacification of their raging volcano. In Icelandic lore, the name by which a thing was called had some significance in determining its actions. The fiery cone which has reared itself above the fissure on Heimaey is now being tentatively referred to as Kirkjufell, the Church Hill.

Part II

BIOLOGICAL SCIENCE EVENTS

Missing in Action: A Living Fossil

(Extinction of the Rare Kouprey Ox: Cambodia, 1970)

The probable extinction of the Cambodian kouprey, a rare genus of ox, in the midst of its native habitat, is perhaps an instance of the ultimate human pollutant. The ironies of fate to which this living fossil is doomed appear to be purely man-made and without reprieve.

The Cambodian forest ox, *Novibos sauveli,* is one of the most ancient of the known bovids and represents a scientific landmark of genetic heritage of exceptional importance. It inhabits a limited area of the northeastern forests of Cambodia and, as of June 1970, some still were believed to survive there—perhaps 30 to 70 specimens known to be left in the *world*. Because of its unique value for genetic studies, it is in the interest of the entire world to safeguard this species. If it becomes extinct, this "gene-bank" developed over a hundred thousand years will be lost *forever*.

In 1937 the first Cambodian ox was described by zoologists in notation, when a living calf shipped from Saigon to the Vincennes Zoo in Paris, France. Although recognizably different from any other wild cattle, it was described in rather superficial terms as a new subspecies of a bovid and named Vibos sauveli. Incidentally, this was the only kouprey ever known to have been in capitivity for any length of time. Ironically, it died of starvation early in World War II.

In 1940, at Harvard's Museum of Comparative Zoology in the United States, it was classified as belonging to a new genus, based on a study of the skin and skeleton of an adult male sent to them. This reclassification was justified by important differences between the kouprey and other forms of wild bovids. Essentially, however, the animal was determined to be of major importance because it is extremely primitive in many of its anatomical features and probably the nearest thing to a relic of the possible missing ancestors of the Indian hump cattle—thus making it doubly important to have at least one and preferably two herds in captivity for detailed scientific study.[1]

In 1952 a documentary film of the wild cattle of Cambodia was made and the results of the ecological study of the kouprey carried on at the same time were published five years later by the Institute of Science and Technology, Philippines. Yet no living specimens were taken into captivity and none are known to be anywhere in the world today.

Knowing this rare species was to be found only in their country, the Cambodians have always taken great pride in them and have in the past made extraordinary efforts to insure their protection. Some years ago the Cambodian government set up a reserve for that purpose and issued postage stamps bearing the kouprey's picture. They also co-operated with the U. S. National Academy of Sciences on an expedition that was supposed to capture 3 or 4 kouprey to establish a captive herd.

Since there were none of the species in captivity on which to test the tranquilizers, a basic nicotine drug which had worked well with cattle was used. Unfortunately, however, the four animals tranquilized and captured had an allergy to these drugs and none lived more than fourteen hours.

In mid-1970, reports showed that the areas where the kouprey was known to survive were limited to three locations. However, two of these probably no longer have any kouprey left in them. One was a reserve in the northeastern corner of Cambodia, known as the Lomphat area, which probably contained 20 to 30 kouprey. But Lomphat has also been the headquarters for Vietcong military activities for some time and there is little hope that any animals are left there. Because beef is one of the Vietcong's favorite foods, and because the supply situation is extremely difficult, it is almost certain that with the readily available live ammunition they have been supplementing their diet with wild game.

The second location known to have been inhabited by kouprey was a fauna reserve at Phnom Prich, south of Lomphat and very close to the Vietnam border. When last visited, there were not more than twenty animals there. Since then, it has been so overrun by the military that there is little hope any of the animals still survive. If they do,

[1] One of the interesting features of the kouprey, which helps to substantiate its designation as a relic form of a living fossil, is that the molars of the specimen studied at Harvard are most like those of *Proleptarbos*, an extinct fossil ox.

it is more than likely that they have been so widely scattered that they cannot find each other for breeding purposes. But even this remote possibility grows dimmer with recent reports that, in addition to being heavily bombed, burned, and defoliated, a long-lasting noxious gas was also released to the area by withdrawing American troops.

The only remaining group of kouprey, if any remains, dwells in the fauna reserve of Koulen Promtet on the Laotian border. This most important concentration of kouprey—about 40 were located there in 1969, before the rains—provides the best hope of the survival of the species and the preservation for genetic purposes and future studies. But this large reserve of 1,467,000 hectares in northern Cambodia lies between the town of Stung Treng and the Laotian border in the Khet Preah Vhar province and latest reports indicate that the reserve is overrun with Vietcong who doubtless are shooting kouprey, wild boar, and banteng—all of which were at one time plentiful inside the reserve. In fact, zoologists are fearful that the Cambodian Army, once alerted to the value of a kouprey calf or even ordered to capture one, might recklessly machine-gun all the females in an attempt to take their young.

So it seems that the unfortunate kouprey has never lacked for human predators. Not long ago, it was reported that three members of the international big game fraternity were anxious to add kouprey to their private collections. Since they had never shot the species, the scion of an old socially and financially prominent American family, the brother of a near-eastern ruler, and a French diplomat were offering $10,000 to the Cambodian government for the privilege of entering the reserve to shoot two or three of them. Their "offer" was politely declined by the government out of concern for their safety because of current military activity there.

City in a Web

(St. Louis Spider Invasion: St. Louis, Missouri, U.S.A., 1969)

When unusual flying or floating material was first observed in the air near the St. Louis, Missouri, airport, people thought it was just another more intense condition of the recent air pollution in their city. Immediate speculation had it that the masses of fluffy, sticky, thread-like particles drifting northward across the city were a synthetic precipitate of the thick and heavily polluted air. However, the mystery deepened as dime-size to 10-foot-long globules were seen—some falling to the ground—on the afternoon of October 8, 1969.

The riddle was solved and man absolved, for once, when the county health department division of air pollution control described the umbrella cloud over St. Louis as a gigantic spider web. Actually, the sinister "science fiction" substance in the air was identified as webbing cast loose by a mass migration of ballooning spiders. These spiders weave treetop webs into which they discharge their eggs. Then, severing the webs, the spiders release their babies, "cradle and all," to be dispersed by the wind.

Weather conditions in St. Louis on the afternoon of October 8, 1969, apparently were such that enormous amounts of webbing descended on the unsuspecting human animals caught below.

Strangely, all but one of the perpetrators escaped and this lone specimen was never accurately identified by the city health department. They disappeared as suddenly as they had come and with them all traces of their unexpected arrival.

Just one year later the residents of St. Louis were visited by one of nature's more pleasant phenomena, the invasion of butterflies by the billion, an ample reward, it seemed, for this year's sticky little joke.

Butterflies by the Billion

(Monarch Butterfly Migration and Peak: Canada, Midwestern North America, Central America, 1970, 1971)

A bright curtain of color wafted gently over St. Louis, Missouri, in the fall of 1970. It lingered, shimmering soundlessly overhead and fluttering quietly in the trees, drifted on, vanishing as mysteriously as it had appeared, leaving the atmosphere a little duller for a memory of startling beauty.

Like the delicate grace notes of color in a Dufy painting, this living panorama was created by literally millions of monarch butterflies (*Danaus plexippus*) migrating from the northern Midwestern states and the lakehead area of Canada to the Gulf states and Mexico. In St. Louis, brilliant clouds of thousands were observed, with hundreds daintily roosting in a single tree; while enormous quantities of these butterflies were seen in most of the Midwestern states.

The North American monarch is distinctive in its coloration and markings. The overside bears a black-bordered triangular wing which is deep on the trailing and outer edge and sprinkled with white oval polka dots. The inner portion of the wing is of a deep apricot color and is segmented by lines. The underside has a similar border but the inner portion is deep yellow and segmented by black lines like a Tiffany glass design.

Many lepidoptera are famous migrants; however, the North American monarch is the only species that every year demonstrates a unique feat of strength and endurance, hardly in keeping with its fragile appearance. These lovely but hardy creatures annually perform a two-way migration—with the same individuals flying southward in the autumn and northward in the spring. They have even crossed the Pacific Ocean, colonizing Hawaii and Australia, and have occasionally been known to reach Europe.

Although these migrations are expected annual occurrences, one of this magnitude has not taken place in eight years. Large migrations are usually observed cyclically about every six years, with population levels remaining extremely low during the interim. The levels rise substantially during the subsequent two-year period, after which they drop off again.

Two interesting questions still surround these migrations: what causes the pronounced fluctuations in population level and why does the migration begin when the temperature is still high and food abundant?

Two possible answers and an interesting related observation are suggested by the scientists studying them.

During the peak years the butterflies seem to be resistant to a specific virus infection from which it is impossible to protect them even under the most sterile laboratory conditions during the "off" years.

As for the time of migration, which usually begins in mid-August, it is possible that the decrease in day length (photo-period) is a factor. The last generation of butterflies, under conditions of reduced photoperiod, showed underdeveloped gonads, a high fat content, and increased motor activity (flight). Upon termination of the migration, the fat is used up and the monarchs mature sexually.

Since 1970 was a pre-peak year, said the scientists, the numbers of butterflies to be seen in migration in 1971 might very well be in the billions.

The peak population occurred as forecast. Throughout most of North America during the summer of 1971, "unprecedented numbers" of monarchs were cheerfully reported. Thus, once again, nature has demonstrated her infinite variety and whimsy. As for the people of St. Louis, who have learned to take the bitter with the sweet—that is, the spider webs with the butterflies—theirs is an attitude of guarded curiosity for the future.

Like a Moving Carpet

(The Victorian Mouse Plague: Australia, 1969)

In early 1969 the inhabitants of Victoria, Australia, began to observe startlingly large numbers of the species *Mus musculus,* the common house mouse. By November and December 1969 the numbers had reached plague proportions in the Mallee areas, where the farmers estimated their numbers to be in the millions.

The Mallee region is one fifth the area of the state of Victoria, covering approximately 172 square miles and encompassing 11 million acres. By the end of the year the mice there were said to be "like a moving carpet," with an estimated density of more than 200 per acre.

No "wee cowrin' trimorous beasties" were they—and, bereft of a parti-colored magician to pipe them away, the farmers trapped as many as 300 to 400 a night in order to protect their crops.

The plague extended throughout the whole of the Mallee area and into some of the Western District. As a matter of fact, it undoubtedly spread more widely than that, as many other areas were not inspected. Oaten haystacks near Maryborough, Lismore, and Donald were ravaged by mice, and crops, including rice, were severely damaged in New South Wales and South Australia where the mice appeared to have been present for about the same length of time as those in Mallee.

Surprising as it may seem, mouse plagues have been previously reported in the twentieth century—in 1917, 1932, and 1949. However, comparisons are not possible because the methods used in judging their numbers and the damage done are unknown. Nevertheless, two features of the most recent infestation appear unique if somewhat disquieting. December 1969 was the first time mice have ever been reported attacking growing rice and seed crops, while the second unsettling feature was the report that none of the methods used gave complete control.

The genus Mus, occurring naturally on all major land masses, has been distributed by man to all inhabited areas of the earth and has become naturalized. House mice reach maturity early and are usually ready to mate two to three months after birth. Gestation runs two to three weeks, occasionally less, and litters generally contain one to eighteen pups depending on the species. They are known to consume a variety of foods—grains, seeds, arthropoda, vegetation, even flesh if available—and in turn embody the major prey group of most large carnivorous mammals, reptiles, and birds.

An interesting and perhaps related sidelight on the social habits of the common house mouse is the recorded abnormal behavior pattern and metabolic disturbance that occur when a population explosion leads to overcrowded living conditions. The resultant "psychosis" then often produces a suicidal migration similar to the lemmings' march to the sea—thus restoring nature's delicate balance within the species.

Within the larger framework of ecology, of course, tipping the scales experimentally by decimating large populations of mice in North America resulted in the expected sharp increase in fox and coyote predation on poultry and sheep.

However, it is doubtful that the Victorian mouse plague of 1969–70 was in fact any migration of mice. Scientists, who admit "little is known about the common house mouse when it is living in the field and even less is known about its control on a field scale," claim the plague was simply a build-up of the regular low mouse population in the area. The extent and success of the build-up and spread to plague proportions would depend on such factors as previous rainfall history, the soil type (for easy burrowing), weather temperatures, and especially the available food supply and the protection afforded the food.

The scientists investigating the phenomenon tell us that it is possible that unseasonal rains early in 1969 gave rise to an early growth of food which produced a slight increase in the normal mouse populations at that time of the year. Then, having survived the winter in sizable quantity, the mice were present in above normal numbers in late spring and early summer to feed on ripened grain in the ear and on the grain which had fallen to the ground. This caused an additional increase in their numbers.

The soil was sandy and suitable for burrowing and weather conditions were also satisfactory, enabling the by now oversubstantial mouse population to feed on the grain stored in the paddocks until room was available in grain silos, on grain escaping from some bulkhead wheat storages which were not mouseproof, and on other appropriate mouse food such as haystacks in paddocks.

The number of mice at these food sties was very high while there were few in areas away from the food and none at all to be seen in the daytime. Since in towns the numbers, although higher than usual, could not be classed in plague proportions and night traverses of roads showed a low population of approximately two mice per mile, the scientific observers ruled as unlikely any suicidal migration of the house mouse.

Since estimating the damage and destruction caused by the infestation is a measure of the intensity of the plague, it is interesting to note that grain in stubble paddocks was eaten by the mice to such an extent that there was no stubble grazing for sheep in the autumn. There was a considerable loss of oaten and wheaten hay conserved for fodder, and over-quota wheat stored in paddocks was also subject to much wastage. Economic damage to rice, sorghum, and maize, in the Riverina was estimated at 20 per cent, with irrigated properties adjacent to dry areas most affected and mice more plentiful in the drier regions. As mentioned above, this was apparently the first time mice were recorded attacking growing rice and seed crops; and in addition to damage to the farmer's storages, many of their fruit and vegetable crops were also devoured.

The impulsive and widespread use of poisons by local inhabitants soon proved ineffective in controlling the mice but capable of creating a serious anxiety in the town for the well-being of their domestic stock. Among the poisons used were such banal compounds as strychnine, phosgene, organophosphates, and parathion. Even chlorinated hydrocarbons, e.g., Endrin, were used with the fungicide TMTD used as a repellent—but these, too, were resisted by *Mus musculus* and two weeks before Easter 1970 the mouse problem was reported acute and control measures by means of poisons were investigated by the Vermin and Noxious Weed Destruction Board. Although none of the methods used had been found to give complete control, field use did aid control in situations such as the heavily hit bulkhead storages which were enclosed by a fence and then fumigated. The result of this procedure was merely a reduction of the number of mice in these areas. An unexpected observation, however, disclosed that little damage occurred in storage areas where polythene was used.

One last aspect of this short-lived phenomenon to interest the scientists reporting on it was the heavy publicity it received. Speculating on this fascinating subject with their usual academic detachment, they have suggested the following theories: No. 2—A general interest in the economic structure of the primary industries in Australia; No. 3—Substantial losses incurred by some landholders whose summer-growing coarse grain crops and rice were attacked (mice had not previously attacked rice in the Murrumbridge Irrigation area); No. 4—The main economic loss occurred to landholders whose major acreage comprised summer-growing grain crops in the agricultural regions of eastern Australia—because of recent restrictions on the growing of wheat.

But listed as No. 1 and the most important reason for its unprecedented notoriety was "the advent of television."

Sweden's Black Christmas

(Southwest Sweden Snow Pollution: Sweden, 1970)

On Christmas night, 1969, black snow fell silently and eerily in southwest Sweden, covering 10,000 square kilometers and transforming the bright and dazzling winter scene of southwestern Lake Vättern into a weird and mourning landscape.

Freezing rain accompanied by a grim southeast wind on December 26 and 27 completed this gruesome weather episode as the snow cover became grayish and eventually stark black in some areas with a deposit so oily and slick it was extremely difficult to remove even with detergents.

Animals added to the macabre photo-negative effect by leaving white tracks on the black fields.

Preliminary analyses at the ecological center at Lund showed that both DDT and PCB were present in the deposit though in very small amounts—DDT: 30mg/km^2, PCB: 35 mg/km^2.

The scientific explanation of the mechanism stated simply that snow or polluted snow cleaned the air and assembled the DDT and PCB. However, the Swedish residents of that area will long remember the stark lesson taught by nature in scrubbing its own atmosphere, purifying the air and leaving the residue pollutants to be swept under the carpet by her tenants.

The Toads Joined In

(The Malaysian Frog War: Malay Peninsula, 1970)

The war broke out on November 7, 1970, on a rubber plantation in Malaysia about 160 miles north of Kuala Lumpur. For nearly a week the battle raged fiercely between armies totaling perhaps 10,000. The battle site was a large, shallow, temporary mud puddle (5 to 6 yards across) formed by heavy rains near the Sri Marriammal Hindu temple in the seaport Sungei Siput and the combatants were frogs of perhaps ten or more species.

Early reports dispatched to the press described two armies, one of frogs of various shades of green and the other of brownish, yellowish, and light black hues. Based on descriptions of (local) residents and innocent bystanders, the largest number of frogs were classified as belonging to the following three species common to Malaysia: *Rana chalconota, Rana erythraea,* and *Racophorus leucomystax.*

The frogs ripped and tore at each other ferociously, the reports continued, calling their generals by name, and even carrying off their dead when they could according to local legend. Although observers claimed that relatively few were dead, the casualties were reported to be in the hundreds.

By the time scientists from the University of Malaysia's Zoology Department arrived at the battleground, peace had been declared and the troops withdrawn. However, they did find that the mud puddle now contained tadpoles of the Ranid and Racophorus species. This, then, forced them to a reinterpretation of the previous reports by the local populace—in short they felt almost certainly that the Great Frog War had been in actuality a mammoth Frog Love-in. It seems the heat and frenzy of this mass mating and breeding orgy is typical of frogs that have had to lie buried for some time to survive a drought and then are suddenly liberated by the first heavy rains. The enthusiasm with which they took to mating and their accompanying noises attracted other frogs and occasional toads, who happily joined in.

Several theories have been advanced for the killings. One is that frogs do fight and some are killed during a breeding aggregation of the same species when a male will try to chase away other males in order to have a female all to himself. They jump at each other and in so doing kill with the abrasive structures on their legs and chests.

Another more probable explanation, indicated by the species of tadpoles and eggs found later in the puddle, is that the visiting toads also engaged in the love fest and, as the excitement of the continued climbing and mounting process intensified, emitted certain poisonous secretions from their skins which killed some of the frogs.

Although it seems most of the participants were frogs of the smooth-skinned *Rana* genus, which includes the common American bullfrog, there were also many toads of the genus *Bufo,* to which most American toads belong. The latter are rough-skinned, tailless amphibians with a body size ranging from ¾ inch to 9 inches. Their outstanding physical features include a thick and often warty skin on the back which is generally colored a mottled brown. The underside is smooth and is sometimes unpigmented though males usually are dark-throated. The parotid glands, located in two raised areas behind the eyes, and much of the dorsal skin and its warts contain a poisonous secretion. This poison, which may be ejected when the toad is attacked, is a highly effective deterrent to predators and molesters. It irritates the eyes and mucous membranes and has been known to be fatal to dogs though completely harmless to some creatures. For example, toads are almost the sole diet of the North American hognose snake. And the handling of toads does not cause warts on human skin!

The poison is a milky secretion from which several substances have been isolated and identified: bufagin, which has properties similar to digitalis; bufotenine, a hallucinogenic drug resembling LSD in its effects; and serotonin, which constricts the blood vessels.[1]

Toads are mainly nocturnal and terrestrial. They feed on insects or any small animal they can catch on their sticky tongues and tend to remain in small feeding areas. During the adult phase of their life span they are inactive only in winter and periods of summer drought when they retire to their burrows.

Breeding generally takes place in slowly moving or standing water

[1] Although the Chinese for centuries have used toad poison to treat various medical ailments, we do not since most of the chemically important substances can be obtained from better sources or synthesized.

and may involve mile-long migrations. Six hundred to 30,000 small dark eggs are deposited in two long jelly tubes from which non-descript tadpoles hatch in a few days.

According to the Museum of Natural History in New York City, breeding frenzies are not uncommon following an extended drought, although the large number of species reported to have joined the melee in Malaysia was highly unusual.

To many Malaysians, the frog "wars" have always had a more mysterious and disturbing significance. To them, these spectacular events portend major national disasters. For example, they point out that just twelve days (May 13, 1970) after a frog war was reported at Butterworth, Malaya, the Malay-Chinese race riots began.

The Malaysian Frog War of Sungei Siput began on November 7, 1970, and continued off and on until Friday, November 13, 1970. On that day the Bay of Bengal tidal waves were reported.

A Menace of Birds

(The Scotland Neck Blackbird Irruption: North Carolina, U.S.A., 1969–70)

A population of blackbirds grew to "menace" proportions over the past three to five years near the town of Scotland Neck, North Carolina. During 1970 an estimated 4 to 6 million cowbirds, grackles, and starlings roosted in a 40-acre tract of forested land (mixed pine and hardwood) adjoining the town of Scotland Neck and a nearby housing area.

Among the problems created by the blackbird irruption was the exceptional amount of noise they made at sunup just before departing for the fields where they fed and at sundown just after returning to roost for the night.

Droppings of the birds had a vile odor and prevented people who lived near the roost from using their yards. Another major aspect of pollution was the presence of droppings in a canal that ran by the corporate lands of the town, the housing area, and the roost. Health officials compared the amount of canal pollution to open sewage. It had a high nitrogen content and was coffee-colored.

Birds were dropping dead on and around houses, causing problems with bird lice, and from the health standpoint, there were still other factors. Once the bird droppings became dry and powdery, disturbances caused the powder to be carried by air currents. Though it was considered unlikely, the powder could cause histoplasmosis, a viral lung infection, in humans. But when soil samples were collected and analyzed for histoplasmosis, they were found negative.

However, a real threat of encephalitis had to be considered if the birds left to nest in the spring and returned when there was a sizable population of mosquitoes.

A last complaint came from the farmers in the area who were observing serious bird depredation on their peanut and small grain crops.

According to official state estimates, there were approximately 3 million blackbirds and starlings in the Scotland Neck, North Carolina, roost prior to a harassment project that began on March 18 at a cost of $1,859.23.

Among the various harassment techniques used were spraying the birds with detergent, playing recorded starling distress calls, and deploying various explosive devices, for example, shotguns, firecrackers, etc.

To spray the birds with detergent, Tiergitol 15-5-9 was dropped on the roost from an airplane. When mixed with water, as in a rainstorm, it removes natural oils from the birds' wings and prevents them from flying. The grounded birds are then susceptible to cold weather and die.

The harassment kept the birds from returning to the roost at the end of the day though it was expected they would probably fly to another area to roost or even at that time of year begin their northerly migration. By the weekend, only 10,000 birds remained in the Scotland Neck roost; but within a few days the number had increased to 50,000.

It is known that there are approximately 200 major blackbird and starling roosts—a roost consisting of 1 million or more birds—in the United States, primarily in the Southeast. Most of the roosts occur in unpopulated areas, but occasionally they are found near communities and may, in these cases, present nuisance and/or health problems. Although the over-all blackbird and starling population of the country (about half a billion birds) remains the same, individual roosts begin to wax and wane in size. Since northerly migrations were beginning, harassment would seem to be an effective method of ridding a populated area of the birds; nevertheless the expectation there was that most would return to the same area the following winter, though the site of the roost might shift by several miles.

The forecast was regrettably correct. The blackbirds returned to roost in the same place that they had the previous winter and the harassment of the previous spring to chase them from their roosts and speed their migration was not a permanent cure.

Amateurs estimated the size of the returned population to be around 12 million; but official estimates put the figure between 6 and 8 million, or twice the number of blackbirds and starlings roosting there the year before. The blackbirds began to arrive during the first part of October and reached their peak of 6 to 8 million by the first of December.

This time, exasperated townspeople tried to smoke the birds out of their roost at night. However, none of the methods used to drive birds from their roosts on a permanent basis seem to be effective. The birds usually return to exactly the same area the following year. Therefore, it is currently believed that the best way to rid a specific area of birds, where feasible, is to chop down the trees. Exactly where the birds go when they are chased from an area is not known. It often appears that they tend to disperse.

From September 1970 through March 1971, the bird population at the Scotland Neck roost peaked at about 6 million. Federal and state agencies were again requested to assist in combating the birds.

Although the North Carolina Department of Conservation and Development agreed to a plan of cutting down most vegetation, landowners were reluctant to allow any loss of their trees. In late March 1971 the roost was vacated but the birds returned again in about the same numbers in the fall and winter of 1971–72, and to more or less the same area.

Scotland Neck has been designated as one of the national test sites for experiments in the management and control of pest birds.

Giant Triton versus the "Crown of Thorns"

(Indo-Pacific Starfish Plague: Indian and Pacific Oceans, 1969–70)

To children and the average summer beachcomber, starfish are an attractive, even glamorous and artistic form of sea life. Somehow one does not think of them in terms of their destructive abilities but rather as the more decorative and pleasantly unassuming members of the sea family.

Sea stars, the most familiar class of starfish, nevertheless possess the sinister ability to crawl over any surface and slither through incredibly narrow crevices at a rate of six inches a minute. Despite the deceptively delicate and fragile appearance of their tapering arms, these lovely creatures, which are partial to a diet of oysters and mussels, are also capable of a peculiarly vicious attack. By affixing the suckers of one or two arms on one side of the shell and those of the opposite arms on the other side of the shell and exerting a prolonged and steady pull as it straightens out its rays, the starfish forces the bivalve (oyster, mussel, etc.) to open its shell. The sea star then feeds its voracious appetite by extruding its stomach through its mouth, digesting the oyster, and withdrawing its stomach after the meal is finished. It is possible for a young sea star less than 3/8 inch across to eat more than 50 young clams of half that length in six days. They vary in size from 1/2 inch to over 3 feet across, become sexually mature in less than a year, and produce thousands of young. Consequently, their damage to oyster beds can assume astounding proportions—in the past estimated at nearly $1 million in one state alone. In turn, gulls, ravens, crows, foxes, and various fish feed on the sea stars to protect the balance of nature in our environment.

A somewhat less recognizable class of starfish, *Acanthaster planci,* however, has recently made headlines in even more menacing fashion. This large, ugly, multi-armed variety, commonly called the "crown of thorns" starfish, appears to feed almost solely on coral polyps—its predation primarily limited to madreporite or hard coral.

In May 1969 news of the irruption of *A. planci* on the Great Barrier Reef of Australia was reported. During the next month it was learned that serious infestations were also extant on coral reefs of certain islands in the South Pacific U. S. Trust Territory. Huge numbers of the starfish were devouring the tiny shellfish that create the great coral reefs of the Pacific Ocean at a devastating rate, thereby threatening the archipelagoes.

It appeared that, for the past five to ten years, increasing herds of the "crown of thorns" have been denuding extensive areas of coral of their living shield, exposing them to the continuous erosive action of the sea. As the coral is killed, the reef itself slowly dies and the marine life populating it—often a staple of the island's economy and diet—is driven away. The shore line of the inhabited islands is subsequently exposed to the erosion, flooding, and other hazards of a pounding surf once held at bay by the protective reefs.

In 1966 an Australian study of the cause and extent of damage done by the starfish infestations on certain reefs of the Great Barrier Reef was begun. It was completed two years later and the report was released for publication on August 1, 1969. The findings of this study were little more than astonishing.

The cause of *Acanthaster planci* overpopulation on the coral reefs was attributed directly to the absence of the giant triton, *Charonia tritonis. C. tritonis* is a univalve sea mollusk with a lovely large spiral marine shell. It is named after the mythological Greek god Triton, a merman and son of Poseidon and Amphitrite, who according to legend dwelt with his parents in a golden palace in the depths of the sea. He was most often represented as human down to the waist with the tail of a fish and known to be the possessor of a violent and jealous nature. His special attribute was a twisted shell on which he blew as a trumpet to raise or calm the waves.

The giant triton is also a major predator of the starfish and the only known predator of older juveniles and adults. But it appears the species American Tourist and Shell Collector, lured by the siren song of cut-rate economy flights to these previously inaccessible parts of the world, have been innocent predators themselves in their demand for the pretty twisted conch shells as souvenirs. Thus, the study pointed to a chain of events beginning with the population depletion of the giant triton conch, which then produced the population explosion of the starfish which has resulted in the gradual devastation of the coral reefs.

The study also showed that massive destruction of living corals due to the depredations of the "crown of thorns" starfish has already occurred on most of the inner patch reefs and fringing reefs of the section of the Great Barrier Reef bounded by the latitude of Cooktown to the north and that of Townsville to the south, with only the outer reefs near the 100-fathom contour escaping infestation by *A. planci.* In summary, the report claimed that more than 90 per cent of the corals on many of the infested reefs had been killed by the starfish, with the destruction continuing in many other reefs in the area.

In July and August 1969 a second major study of *A. planci* destruction of coral was undertaken by United States scientists in the South Pacific United States Trust Territory. Previously extensive observation of starfish damage to reefs surrounding the island of Guam had been made. Now ten field teams surveyed sixteen other islands in the area while in a companion effort five other teams inspected the Hawaiian Islands, Johnston Island, Majure, Arno, Kwajalein, and Midway. The findings of the study were released by the Department of the Interior in December 1969.

Of the nineteen trust territory islands surveyed, ten were reported to be clearly infested, showing extensive coral damage and exceedingly large populations of *A. planci* in either developmental states or already formed massive migrating fronts. Three islands had populations large enough to be producing areas of dead coral but could be in a process of either development or decline. The remaining six islands investigated were clearly not infested.

The U.S. study also significantly concluded that reduced populations of the giant triton conch appeared to be one of the causes of the present starfish infestation in Micronesia. In addition, however, the same study pointed accusingly at man for other contributing factors to this ecological imbalance—viz., blasting and dredging and chemical pollution. Both reports recommended the continuation of a comprehensive study of the problem and that immediate steps be taken to eliminate the starfish from areas of infestation.

The mammoth task of eliminating the invading starfish, however, soon assumed Herculean proportions. Like the nine-headed Hydra, which grew two new heads in place of any one slain by Hercules, the starfish merely regenerate when cut up. None the less daunted, man set out to restore nature's balance in his own inimitable if ponderous fashion. On September 28, 1970, fifteen Navy divers were deployed to eradicate more than 10,000 starfish on the reef off Molokai. Each diver carried two hot water bottles filled with ammonium hydroxide and a syringe equipped with a spring-loaded plunger, which permitted automatic refilling from the hot water bottles. Using the equipment in highly infested areas, the divers injected ammonium hydroxide into each starfish, killing as many as 400 during a twenty-minute dive.

Two months later the "crown of thorns" starfish were reported present on coral reefs in the Papua and New Guinea areas of Port Moresby, Samurai, Wewak, Manus Island, and Finschlafen. One reef of New Ireland was reported to have been destroyed and, in a bulletin published by the New Guinea Highland Farmers and Settlers Association, concern was expressed for the starfish threat to the fishing grounds of island villages as well as their growing tourist industry.

However, in April of 1970, a new theme had been introduced in this underwater fugue as a population of "crown of thorns" starfish were discovered off the southwestern Pacific coast of Panama. This was interpreted not as a new invasion but rather as an extant population in an underexplored area. A survey of coral around three islands in the area as well as around the mainland determined *A. planci* maximum population densities of 2.2 to 2.5 individuals/100^2m. The starfish were observed feeding on seven coral species with estimates of the feeding rate ranging from 0.22 to 4.02 m. of reef per individual per month. Interestingly, coral growth in areas of high starfish density was *not* demonstrably less than in other areas where the predator was rare or absent. Moreover, quantitative measurements of live to dead ratios of corals in and outside of areas occupied by the "crown of thorns" starfish failed to show any signs of depredation that could be due solely to them.

The report ends on a cool and compelling note with the following statement: "A single individual of Hymenocera species was observed and photographed clinging to the aboral surface of a moving starfish at the Contreras Island. Though the brief sighting gave the impression of symbiosis, this shrimp is thought to be possibly an important predator of the 'crown of thorns' starfish."

Appalachia's Latest Poverty Victims

(The Appalachian Squirrel Migration: Southeastern U.S.A., 1968)

During August and September 1968, hordes of squirrels were reported dancing on the highways, swimming lakes, and damaging farmers' crops. There were newspaper accounts of vast mass migrations, mass starvation, and the presence of squirrels where they had never been before. This unusual activity on the part of the common gray squirrel (*Sciurus carolinensis*) in 1968 was reported throughout portions of the eastern seaboard from Vermont to Georgia by sportsmen, motorists, farmers, game wardens, and biologists.

Thinking the activity was due in some way to a shortage of food, radio and television stations advised people to feed the squirrels and signs in supermarkets and elsewhere gave similar advice in the North Carolina area. Some of the concerned and tender-hearted residents immediately responded with checks to the North Carolina Game and Fish Department to help pay for squirrel food and demanded the immediate organization of a squirrel-feeding program. Enterprising Tennesseans recruited willing Florida residents to gather acorns for shipment to Tennessee. Sportsmen requested an early open season for squirrels with double bag limits, and, indeed, the Tennessee Game and Fish Commission acquiesced.

The New York *Times* reported that "squirrels are Appalachia's latest poverty victims . . . suffering from a severe lack of natural food amounting almost to famine conditions . . . tens of thousands killed daily by automobiles while crossing highways. . . ."

Reports came from everywhere—tabulating highway squirrel deaths at 1 to 5 per mile on the Massachusetts Turnpike, at 100 dead specimens on a 27-mile stretch of highway near Asheville, North Carolina, and, in addition, more dead foxes than usual were noted on the highways—probably because they were attracted to the squirrel carcasses and were also hit by cars.

Since road kills are often indicators of mammalian activity, it was

1. An "igloo" of molten lava—A lava dome forty feet high fountains from the east rift of perennially active Kilauea Volcano on the island of Hawaii during its 1970 eruption. *Photo: U. S. Geological Survey, Washington, D.C.*

2. Kovachi's birth pangs—Flaming, fuming and discoloration of the waters of the South Pacific with tephra and other volcanic matter accompany the attempt of Kovachi to establish itself as an island. It did so, temporarily as in the past, and then subsided beneath the surface once again, owing to rapid erosion. *Photo: Dr. R. B. M. Thompson, Geological Survey Department, Guadalcanal, British Solomon Islands.*

3 and 4. Hekla by night and day—One of Iceland's deadliest volcanoes, Hekla, during its 1970 eruption, which lasted two months. Lethal fluorine gas, with which Hekla's tephra is mixed, killed thousands of head of domestic cattle during this outbreak. *Photo: Robert Citron, Smithsonian Institution Center for Short-Lived Phenomena, Cambridge, Massachusetts.*

5. A modern Pompeii in slow motion—Street lamps glow eerily through the darkening fishing village of Vestmannaeyjar, as a somber volcanic cloud hovers overhead and a blanket of tephra slowly builds up during the early stages of the Helgafell eruption in January 1973. *Photo: John Whitman.*

6. Inspiration for a designer—During the peak of their migration by the billions throughout central North America in 1970-71, colorful monarch butterflies create patterns against a background of leaf and vine. *Photo: Dr. Fred A. Urquhart, Department of Zoology, Scarborough College, University of Toronto, West Hill, Ontario, Canada.*

7. Inside the Glacier Cave beneath Lake Linda, Alaska—A wide-angle lens makes the hole through which Lake Linda drained seem much larger than it actually is. But it was sufficient to provide an exit for the water, which escaped into the interior of the glacier. The day after this photograph was taken, this chamber was destroyed by massive breakdown of the ceiling.

The picture was made in natural light except for a few features illuminated by flashlight. Mark Woodward, member of the party exploring the cave, wears an emergency "firefly" strobe light around his neck.

assumed some sort of unusual activity was definitely taking place among the squirrels. But what it was and why it was remained unanswered questions. About this time, population stress was another theory discussed and ultimately discarded.

The squirrels were swimming lakes and rivers, even reservoirs, and deaths by drowning began to be recorded from the Hudson River in New York State to the TVA reservoirs in the North Carolina-Tennessee areas. Rescues began to take place as concerned animal lovers fished these bedraggled little creatures out of the water while several more aggressive types leaped into some boats on the Barkley Lake in Kentucky to the amazement of their occupants. The *Courier-Journal* of Louisville, Kentucky, gave the following account: "A boatman on Old Hickory Lake near Nashville saw a squirrel swimming far out in the lake. He was so surprised to see the little varmint fight the water, that he threw him a life preserver. Not being one to refuse a helping hand, the bushy-tail hung on while the fisherman towed the preserver to shore. . . . The squirrel jumped off, shook himself, and quickly disappeared into the trees."

Newspaper accounts of the neurotic and hysterical behavior of the squirrels continued as they began to penetrate buildings in towns and cities and "one desperate individual was discovered in front of Asheville's S and W cafeteria."

The Roanoke *Times* wrote, "The lack of food for squirrels . . . has created such a nutty situation, that wild life officials are shaking their heads in bewilderment."

Both the migration and newspaper reports appeared to reach a peak during the latter part of September. On September 30, 1968, a UPI report read as follows:

SQUIRREL CAUSES BLACKOUT

> *Clarksville, Georgia.* Power company officials said a squirrel, migrating from the mountains to the lowlands, climbed into an insulator, and caused a short that left parts of the city without power for more than an hour. The squirrel was also extinguished.

Early explanations for the largely undirected movements of the "20 million" animals were based on food shortage. The squirrels were thought to be "traveling in all directions looking for food," a situation created by a lack of mast due to an insufficient production of acorns, an overabundant squirrel population causing population stress and anxiety, or a combination of these factors.

However, reports from biologists sent into the field indicated that, even though the nut and acorn crop was not as heavy as usual, there was no shortage of food.

Specimens examined in the laboratory were not emaciated or in poor condition due to intraspecific strife in areas of high population density. Many of them were fat with stomachs containing black cherries and acorn mast.

In short, it was concluded that probably "the bumper mast crop of last year produced two highly successful breeding seasons in July and August 1967 and February and March of 1968. The great influx of squirrels for some reason triggered unusual activity. This extraordinary activity seems to be a widespread phenomenon—and because it occurs at a time of the year when food is abundant and there is no relationship to food conditions, it is puzzling."

Squirrel migrations during the past hundred years have become relatively rare since the cutting of the virgin timber of eastern North America in the late 1800s.

However, September migrations of squirrels have occurred and been recorded before. In 1920 and 1949 they were thought to be due to either fleas or psychological factors, since it was realized even then that "food was in abundance at that time of year."

The gray squirrel does not cache food in the manner of the red squirrel, which stores large amounts of seeds or nuts in tree cavities or elsewhere. Gray squirrels store food but usually gather and bury acorns individually under leaves and in the soil, most often during the month of September.

At this time of year they also take part in something called the "fall reshuffle," when they move about and take up residence in new areas. This apparently is due to the squirrels' search for nuts and acorns to bury during the late summer and early fall. If the supply is inadequate, they move on until they find what they need, bury it, and in so doing find a new area in which to settle.

Nineteen sixty-seven was a year with a bumper acorn crop while 1968 in contrast was an unusually poor year for acorns. Squirrel numbers seemed to be low in 1967 but high in 1968. This increase appeared to be due to the heavy acorn crop, which resulted in low winter mortality and increased reproduction during early 1968.

Since this increased number of squirrels found few acorns and nuts to bury in 1968, they kept moving and quickly burying the few acorns they did find, thus compounding the situation so that significantly more squirrels were moving about much more than usual.

An interesting observation was recorded during the rescue operations of gray squirrels swimming the reservoirs. The rescuers found they were completely unable to persuade the swimming squirrels to turn around and swim back to where they came from. It was thought that the swimming squirrels possibly had their attention fixed on some aspect of the far shore from which they could not be distracted.

As for the "dancing squirrels of the highways"—several explanations were hazarded to account for their slightly hysterical behavior. Some observers noted that, as cars approached, the squirrels on the road would sometimes jump into the air, zigzag back and forth, and in general behave erratically.

One sympathetic traveler felt that the migrating squirrels had traveled so far, the soles of their feet were very thin and therefore the vibrations of approaching cars created a tickling sensation which caused them to leap madly about in the air.

A more likely explanation, suggested by the biologists, was that emigrating squirrels caught out in the middle of the road in an unfamiliar area simply became frightened and confused and, not knowing which way to run, behaved erratically.

Finally, as emotions mounted and donations of nuts and acorns and unsolicited checks and cash continued to pour in, wild life officials once again found that humans are harder to understand and manage than game—as tempting a speculation as the possible increase of mass migrations with the current return of forests.

A Stampede from the Sea

(Florida Beached Whale Mortality: Florida, U.S.A., 1970)

One hundred and fifty false killer whales (*Pseudorca crassidens*), weighing approximately 1,500 pounds each, swam onto beaches along a two-mile stretch of Fort Pierce, North Beach, Florida, on January 11 and 12, 1970. They were 15 to 20 feet in length and 3 to 4 feet in height.

The Department of Natural Resources Marine Patrol responded immediately to the persistent calls of awe-struck Floridians with boats and a valiant effort to haul the whales back into the water. But much to their dismay and perplexity, the whales turned around as soon as they were free and swam back onto the shore faster than the boats could head them off, only to beach themselves once again. Only three didn't reach the beach immediately, and they were found somewhat later on Palm Beach, dead of dehydration. The ones that were still alive, however, were sounding and making a special kind of sonar-type noise.

After an examination by a doctor from the Miami Seaquarium, a second attempt at towing the living animals back into the sea was initiated in co-operation with the state Marine Patrol. Approximately 50 whales were removed from the beaches then and were still in the ocean at the end of the day.

By this time, several theories were being advanced for the stranding of the whales. One suggested that the cold water, caused by unusually cold weather (26 degrees), had forced the bull leader of the herd to lose his way in looking for warmer waters.

Another explanation forwarded was that the whales had strayed into shallow water. It has been scientifically demonstrated that an echo-locating signal such as is used by all cetaceans, whether at a sonic or ultrasonic level, attenuates and disappears if it is aimed toward a shallow, shelving sandy beach area. The coastal topography of the Fort Pierce area has these characteristics as does the coast line of Grassy Key, one of the Florida Keys, where approximately 60 pilot whales (*Globicephalus melaena*) came ashore in 1966.

Another theory for the strandings was the distress of the leader and the infection of the whole herd by an epidemic type of illness. In addition, it is known that whales become disoriented when they are in water too shallow for them to feel safe or navigate properly. Herd instinct may then take over and mass hysteria results in something akin to a cattle stampede.

The blood samples drawn from six of the stranded whales before they were towed out showed no indication of bacterial disease—only signs of kidney or liver dysfunction were found. However, this could have been due to the prolonged period the whales spent out of the water—more than twenty-four hours.

The conclusion drawn was that the pod of beached whales had reacted to a combination of factors involving biological, beach, and tide conditions which might have occurred anywhere in the world but were, nevertheless, highly unusual for the Florida coast.

A Sea of Troubles

(Oil Spills and Animal Kills)

Nature has very carefully segregated its vast store of the liquid mineral known as petroleum from the animal and vegetable kingdoms. Plant and animal organisms, buried deep in the seas of past geologic eras, were sealed off from the air by new mud deposits and subjected to slow decomposition. In this way the petroleum of commerce was formed and, for the most part, confined deep underground.

Man has diligently searched out these reservoirs and brought oil to the surface for his own purposes. In so doing, he has discovered that a little bit of spillage goes a long way toward adding to the pollution of his habitat and the blighting of other living organisms.

There seems to be no end to the rupture of ship hulls, storage tanks, pipelines, and oil wells themselves. And the increase of offshore drilling is eliminating the middleman in the contamination of the seas.

Because oil spills have such a catastrophic effect on the environment and its life forms, they are classified as biological events. Here are only *some* recent examples that tell why.

SANTA BARBARA OIL SPILL, 1969

On January 28, 1969, a well was being drilled five and a half miles seaward of Santa Barbara, California, when at a little more than 3,000 feet below the drilling floor (40 feet above the ocean surface on the platform) a gas deposit was hit. The gas under pressure forced its way up through the drilling hole, bringing crude oil with it. Oil and gas continued to gush from the well for ten and a half days until the flow was temporarily controlled by plugging the well with cement on February 8. By this time Santa Barbara Channel was covered with an oil slick many hundreds of square miles in area; one news agency reported that the area of the slick was 800 square miles.

Estimates of the average daily flow from the well into the ocean during the initial blowout varied from 300 to 500 barrels of oil daily to a conservative estimate of 5,000 barrels of oil spillage per day on the basis of the area and thickness of the oil slick, to a less conservative but still realistic calculation which placed the figure at 16,000 barrels of oil per day.

Oil seepage continued at a very slow rate until February 12, at which time gas under pressure, having forced its way through several fissures beneath the ocean floor and having passed through the ocean floor in several places, opened new and less easily controlled oil and gas leaks. This flow continued unchanged at a rate of 500 barrels a day through late February when there was a brief increase in the seepage. However, during the first week of March the amount of leaking oil was reduced by at least half to 200 barrels a day. This level of contamination, it was believed, continued through May. Again estimates varied, though by conservative calculation it was figured that more than 3 million gallons of oil had spilled into Santa Barbara Channel by the middle of May. In contrast to these estimates were others as low as 15 barrels with pollution seepage (that portion of natural seepage that is contained by collecting devices) less than 10 barrels a day.

In July the leakage was reported variously from 50 to 100 barrels per day. It was thought that the decrease in leakage between May and July may have been due to either reduced pressure from pumping oil out of the reservoir or better collection techniques.

This relatively low level of leakage persisted until shortly before December 16, 1969, when a new 50-square-mile oil slick appeared. On December 17 a leak from a cracked pipe was reported. Pumping was shut down long enough to repair the pipe. During the period of no pumping, it was reported, massive amounts of oil began to seep up from the ocean floor but upon resumption of pumping the seepage diminished. All attempts to contain or completely stop the leakage have proved only partially successful and oil continues to pollute the water and shores of the Santa Barbara Channel.

Since the initial blowout 100 miles of California coast line have been contaminated with oil. The December 16 leak carried new oil pollution to a 12-mile stretch of beach southeast of Santa Barbara in an area where large numbers of birds winter.

Except for the damage to sea and shore birds, the over-all effects of the oil pollution on wild life in the Santa Barbara Channel are largely unknown. Early in February 1969, 20 to 30 to 50 dead birds were washed ashore, though there was no way of knowing how many dead were out at sea. There was no information about the effect on invertebrates or plants. The prognosis for the birds was bad and the report stated: ". . . the birds all die if they settle in the oil for they preen themselves trying to get the oil off their feathers, and they then ingest the oil which is toxic."

The press reported "estimates of numbers of birds that died range from 3,000 to 5,000 and higher." As for the success in cleaning birds during the initial spill, it was reported that it had diminished to just a handful of surviving individuals. Out of 1,500 birds treated for oil damage, 10 or 15 were let loose and it was not known whether they survived.

Sea birds that survived or did not become involved in the initial spill returned to Santa Barbara to winter. Of these, one flock of about 200 grebes were heavily inundated by the December 16 slick.

A number of different species of pelagic and shore birds were contaminated by the oil. Among those that were reported are cormorants, loons, common grebes, eared grebes, California murres, scoters, marbled godwits, willets, dowitchers, western gulls, California gulls, and Heermann's gulls. On February 10 it was reported in the newspapers that most of the birds harmed were loons and western grebes. It was also reported at the same time that Corexit, the chemical most widely used in this incident, which acted as a dispersing agent, causing the oil film to break up into small droplets, which theoretically sink to the bottom, is also a wetting agent like a detergent and is gravely toxic to many forms of marine life.

Although it was difficult to assess damage to the marine environment as a whole, several reports were issued following initial observations. Certain populations of marine invertebrates seemed to be killed outright by the oil while others appeared to suffer no ill effects.

Oil was reported to be causing serious damage in the intertidal zone, especially to the abalone, mussels, and vegetation. In late February 1969 at Rincon Point, California, large numbers of periwinkles and "wavy tops" were found dead, whereas most of the sea anemones were still alive. It was also noticed that the tidepools with oil were empty of nudibranchs and rock crabs. On the beaches of Point Conception normal reactions were observed in sea anemones and seemingly healthy mussel colonies in spite of heavy oil coating. Indeed, fish eggs and larvae in areas of oil contamination appeared to be normal. Surface phytoplankton and microplankton were reduced but there was no convincing evidence that it was due to the oil. Among zooplankton samples, a variety of crustaceans was observed to be active in oil-slick samples. As to the effects on fish populations, several authorities reported that the contamination had no observable influence on fish life.

A few dead sea mammals were found but for the most part they seemed to avoid direct contact with the oil. In the case of whales, it was reported that if they see discolored water they generally skirt it if possible.

In March the death for that month of six California gray whales was confirmed and compared to an average of about one dead whale washed up on the beaches every other year. At that time there was no evidence linking the whale deaths to the oil disaster.

In February a dead bottlenose dolphin was examined and found to have been smothered by oil in its blowhole. But most often reports indicate that the fur seals, elephant seals, and California sea lions on the islands in the Santa Barbara Channel were not damaged by the oil. Tests were carried out for petroleum residues on tissues from several elephant seals and sea lions. The results from all tests were negative.

RHINE RIVER FISH KILL, 1969

Dead fish were noticed in the Rhine River at Bingen, West Germany, beginning on June 19, 1969. A wave of dying fish continued to sweep down the Rhine toward the Netherlands. Dutch authorities were warned of the circumstances on June 23 when poisoned water was crossing the German-Dutch border. On the following day Dutch officials took samples of the water and identified the toxic substance as Endosulfan, a chlorinated cyclic hydrocarbon marketed under the name Thiodan.

The samples contained 0.44 to 0.70 micrograms of Endosulfan per liter of river water. By July 7 levels of Endosulfan were considerably reduced and the contamination was announced to be no longer dangerous.

Official sources estimated a fish mortality of 40 million or approximately 6,000 tons.

The source of Endosulfan, presumably introduced into the Rhine in the area of Bingen, was apparently not traced.

"GIRONDE" OIL SPILL, 1969

A collision between the French oil tanker *Gironde* and the Israeli cargo ship *Harbashan* on August 19, 1969, near the Bay of St. Brieuc destroyed two oil tanks on the *Gironde,* emptying 1,500 tons of fuel oil into the sea. By August 25 strong winds had moved the resulting 7 by 12-kilometer oil slick onto the beach between the two villages of Erquy and Val André on the north coast of the Brittany Peninsula. An area of beach 4 kilometers long by 300 meters wide was contaminated. Powdered chalk and sawdust were used in an attempt to clean up the oil. Although some reports claimed no biological effects, there were accounts to the contrary. One report stated that cormorants and other sea birds were found on the beach and treated for oil damage.

A colony of 3,000 to 5,000 birds of the species *Melanitta nigra* were thought to be in the Bay of St. Brieuc at the time of the spill. Also, parts of the bay are known to be very rich in shellfish, particularly the edible species of cockle, *Cardium edule*. There have been no reports as to whether the oil damaged either the bird colony or the shellfish.

GREAT BRITAIN BIRD MORTALITY, 1969

Initial indications of the sea bird mortality in the northern Irish Sea were noted in the Firth of Clyde in September 1969. Within a few days more than 6,000 dead and dying birds came ashore on its coast in western Scotland. Many of these birds were oiled and oil was at first thought to be the culprit; however, within a few weeks it became clear that a major bird "wreck" was taking place on the coasts of Ireland, England, and Wales as well and that oil was, at least, not the only cause but probably one in a combination of factors.

Of these, the presence of high levels of polychlorinated biphenyl residues has considerable priority though little is known of the physiological action and toxicity of these substances.

The majority of the more than 10,000 dead birds that came ashore were guillemots. Other species affected were cormorants, shags, puffins, razorbills, and gannets.

SULE SKERRY ISLET BIRD KILL, SCOTLAND, 1971

Sule Skerry is a small flat islet some thirty miles west of Orkney and north of Sutherland in the North Atlantic with a huge colony of some 60,000 pairs of puffins, *Fratercula arctica*, among other species. The lightkeepers on this island first reported about 60 birds oiled on April 26, 1971, with more dead in the water later. Since this is the most inaccessible islet in the British Isles, it was concluded that it was impossible to mount even a reconnaissance, let alone a rescue operation. However, an observer passing by in the Fisheries Research Vessel *Explorer* a month later noted there were still a few oiled fulmars and kittiwakes about at sea. "This incident is very serious because Puffins are the seabird which has shown the most marked decline further south in recent years, due in part, to oil pollution."

This pollution affected "our main northern seabirds breeding colonies; hitherto the damage has mainly involved birds further south in the winter quarters. We suspect the oil may come from either N. Sea drilling rigs or tankers washing out off N. Scotland."

KODIAK ISLANDS OIL POLLUTION, 1970

According to a report put out by the Federal Water Quality Administration in May 1970, "At least 1,000 miles of coastline along the eastern side of the Kodiak Islands, Alaska Peninsula, Kenai Peninsula and Montague Island and Prince William Sound were estimated to have been affected by oil. Of 27 beaches surveyed from the Barren Islands to Sitkanek Island, 85% were found either contaminated or exhibited oiled bird carcasses and remains.

"A minimum estimate of 10,000 seabirds were killed. The U. S. Fish and Wildlife Service stated that 100,000 birds may have been killed by the oil. The estimate of 10,000 birds did not account for unobserved losses resulting from predation, disintegration, drowning and sinking of oil, contaminated and weakened birds, nor those blown past Kodiak Island.

"500 marine mammals, mostly hair seals, but including sea lions and sea otter were estimated to have come in contact with the oil. Although no seal or sea lion mortalities were reported in the immediate vicinity of the oil contaminated area, large spots of oil were observed on animals taken by seal hunters. One dead, oil contaminated sea otter was found on a beach near Kodiak, 25 miles from the closest known otter population. Seal hunters observed 55 seals with obvious oil contamination. The oiled furs were cleaned in gasoline and sold. Some furs were completely saturated and cleaning was ineffective. One seal hunter noticed that the hair started to slip after gas and detergent washing. The loss of hides is an economic hardship to the hunter who depends on seal-hunting for income. Fishing gear, including crab floats and lines, were oil-coated." Although "the only direct damage to commercial fisheries was to fishing gear . . . potential damage to commercial fishermen was believed to be significant; however, research was needed to determine the full impact of long-term pollution. Species that could be affected if oil pollution was not abated were: the one million fur seals that migrate through the Gulf of Alaska during the spring; and virtually all species of salmon, crabs, shrimp and bottom fish that were exposed to oil pollution during the critical stages of their life cycles, including pre-larvae, larvae, and juvenile stages."

Although the source of oil damage that started in February 1970 was unknown, possible sources of oil in order of suspicion were "tankship slop oil discharged following tank cleaning, tankship dirty ballast from ships (such as fuel oil), natural seeps, sunken derelict ship breaking up, sunken barge off Marmot Island, land spill from storage or loss during vessel loading, oil platform losses, foreign fishing fleet (Russian) off east coast of Kodiak Island, Naval operations."

After checking into all these possibilities, it was concluded that the oil pollution was caused by deliberate discharges of slop oil and/or oily ballast from tank ships. This dirty ballast and/or slop oil could have been discharged in a legal manner, outside the 50-mile zone. However, while some shipping companies adhere strictly to company rules and federal law, others do not. Therefore, the oil could have been discharged inside the legal limit. It was thus recommended that all U.S. tankships be requested to "cease pumping oil of any concentration enroute between west coast ports and Alaska terminals . . . that Alaska oil loading terminals be required to install treatment units capable of processing the entire ballast received," and that international action be initiated "to prohibit discharge of oil anywhere under the 1954 convention."

SOVIET OIL PIPE LEAK, U.S.S.R., 1971

According to the Moscow correspondent for the New York *Times*, ". . . the pipeline break occurred in an area where the 40-inch line parallels the west bank of the Ural River between the cities of Guryev and Uralsk in northwestern Kazakhstan. The 800-mile pipeline was placed in service last year between the Mangyshlak oil fields on the northeast shore of the Caspian Sea and a refining complex in the Kuibyshev area on the Volga River.

"The transmission of Mangyshlak crude oil posed particular problems because it has an unusually high wax content and tends to solidify at normal temperatures. Heating installations had to be added at regular intervals to keep the oil moving.

"During recent cold weather, according to *Pravda*, the pipeline steel was evidently unable to withstand the strain between the low outside temperature and the high inside temperature from the heated oil and several leaks developed.

". . . Maintenance crews were able to contain the spilled oil behind temporary banks of earth thrown up by bulldozers and other earth-moving equipment. [It was feared that] the embankments of frozen ground mixed with snow would give way as spring approached. Oil was already beginning to seep [out in an area] roughly halfway between Guryev and Uralsk. Water from the melting snow was expected to carry the oil into the Ural River."

This break in one of the Soviet Union's biggest pipelines threatened the sturgeon fishing grounds of the Caspian Sea, a source of caviar, and fertile bottom lands of nearby collective and state farms. The government newspaper *Pravda* stated that the delay of control measures would result in serious damage to that area.

AMER OIL SPILL, NETHERLANDS, 1970

The Zoologisch Laboratorium in Amsterdam reported as follows:

"On December 27, 1970, a supply tank of an electric power station containing 16,000 tons of crude oil suddenly burst open. This tank was located on the southern bank of the Amer River, near Geertruidenberg, Province Noord-Brabant, the Netherlands. It was situated just upstream of the Brabantse Biesbosch, an inland delta area occupying 120 square kilometers and lying between the southern tributaries of the Rhine and Meuse rivers. This region, isolated and with a complex of islands, is of special interest in winter as a wild life refuge and as a result 845 hectares [approximately 2,088 acres] are now set aside as a state nature reserve.

"After the tank ruptured, approximately one half (8,000–9,000 tons) of the oil streamed over the low embankment into the Amer River. Due to the low temperature, the oil was rather dense and floated on the river in a doughy mass of up to half a meter thick. The mass, driven by the eastern wind, floated toward the Biesbosch. Most of the oil mass was removed with five draglines on barges, and by January 6, the authorities claimed that only 400 tons were left in the Amer. Other observations suggest, however, that much more oil remained. Mechanical efforts to cleanse the river of oil had stopped on January 6 and not for some time did further efforts at cleansing resume. On January 25 a contingent of 100 military men began removing the oil by hand.

"In the meantime the thawing of the ice in mid-January aggravated the situation. The thaw had allowed the oil to invade the formerly ice-blocked creeks of the Biesbosch and to flow onto at least half of the stream banks in the area. This heavy contamination of the banks, from one to five meters into the bordering vegetation, remained during the spring and summer. New growth of the reed vegetation began to hide this contamination from view but the sticky material on the bottom still caused fouling.

"On the fourth, fifth, and fifteenth of January and the eighteenth of May 1971, it was stated that all efforts would be made to remove the oil remnants. On the ninth of July, it was stated that these remnants could not be mechanically removed without significantly disturbing the scientific value of the nature reserve. As a result, the oil was to be allowed to degrade at the spot.

"The leakage caused the death of at least 5,000 birds of various species. Due to the difficulty of recovering dead birds, the casualty figures that were assembled were minimal ones. Details of the most seriously affected species are as follows:

Mallard—1,500 to 2,000 casualties comprising the total Amer population.

Pochard—at least 500 casualties.

Tufted Duck—200 to 400 casualties.

Widgeon, Pintail, and Teal—50 for each species.

Graylag Goose—300 individuals (75 per cent of the population) were noticeably contaminated, many fully blackened.

White-fronted Goose, Bean Goose—500 to 600 dead and seriously contaminated.

Bewick's Swan—80 per cent of the 210 individuals present were contaminated and soon the entire population had been killed or chased away.

Gulls (Black-headed and Common)—300 to 400 victims; many others oiled but able to fly.

Coot—all killed (between 200 to 300).

Moor Hen—all of an estimated 150 birds living on the contaminated riverbanks were killed.

"Many other species were badly affected by the oil, but owing to their small numbers, or relative invulnerability, their losses per species were not expected to exceed 25."

"ARROW" OIL SPILL, CANADA, 1970

The oil tanker *Arrow*, carrying 16,000 tons (3.8 million gallons) of Bunker C fuel oil, ran aground on Cerberus Rock in Chedabucto Bay, Nova Scotia, on February 4, 1970. Immediately, at the time of grounding, the forward oil tank bottoms ruptured when the tanker struck the rocks.

Sixty-mile-an-hour winds blowing from the south carried the oil to the north and east where it lodged principally in the area from Kavanaugh Point, on the opposite side of the bay from the town of Arichat, to Cape Auget. This stretch of beach, about three miles long, was heavily inundated with oil. Then the wind changed and for three or four days blew almost entirely from the north. During this period oil was almost constantly released from a variety of leaks in the vessel.

From Black Point, or Fogherty Head, east to George Island, which is beyond the town of Canso, the southern coast line of Chedabucto Bay received a rather heavy load of oil. The northerly wind kept large pools of oil in the water against the south shore. From the shore line, about twelve miles long, there were continuous iridescent, very thin oil slicks streaming off to the north. A long, thin slick from the ship floated eastward with little plumes of oil heading off to the north. As for impact on the shore, these were the only two areas affected. With regard to the effects on wild life, people reported seeing varying quantities of oiled birds. Among those that were identified, the majority seemed to have been merganser, scoter, and eider ducks. With the exception of intertidal life, most marine forms, including phytoplankton, did not seem to have been damaged by the oil, which is relatively non-toxic compared to crude oil or kerosene. There undoubtedly was destruction of life along the intertidal zone. The crabs, limpets, seaweed, etc., were completely covered in oil in this area.

By the fourteenth of February an estimated 1.5 million gallons of Bunker C oil had spilled from the tanker into Chedabucto Bay. Between the thirteenth and sixteenth of February a discontinuous 1.6-mile survey of oil-contaminated coast line in the bay was made in search of dead and oiled waterfowl. A total of 308 dead birds and 41 oiled birds were found and identified according to species as follows: 117 old squaws dead, 27 oiled; 35 red-breasted mergansers dead, 5 oiled; 37 unidentified murres dead, 1 oiled; 24 horned grebes dead, 4 oiled; 5 red-necked grebes dead; 30 unidentified grebes dead; 16 dovekies dead. Dead and oiled waterfowl of the species black guillemot, bufflehead, goldeneye, common loon, scoter, and gull were found in numbers less than 5. In Chedabucto Bay, of the

130 miles of coast line where dead birds could be found, 50 miles were uncontaminated in varying degrees from very light to very heavy at the time of survey. By weighting the above figures, the census takers estimated a total of 2,300 dead birds in the area. They also did several aerial transects over a few days to find out how many live birds were using the total surface of the bay. This would include all additional birds that could be affected. A total of 1,974 birds were observed with the following species distribution: 460 herring gulls, 420 old squaws, 410 common golden eyes, 170 black ducks, 140 red-breasted mergansers, 130 Iceland gulls, 110 great black-backed gulls, 40 ducks unidentified, 40 birds unidentified, 17 scoters unidentified, 7 white-winged scoters, 13 common eiders, 10 red-necked grebes, 1 horned grebe, and 1 unidentified alcid. An example in the survey for dead and oiled birds from one 0.3-mile stretch of the uncontaminated Boulder Shore area showed 1 dead old squaw and a series of partially oiled and actively preening birds (including 13 old squaws, 1 surf scoter, 1 red-breasted merganser, 1 horned grebe). At the same time, 40 goldeneyes and 50 gulls that were uncontaminated and behaving normally were observed. Compared to their total population in the bay, the number of dead and oiled goldeneyes and gulls was much smaller than that of similar comparisons with other species of waterfowl involved in the spill. This may have been an indication that both goldeneyes and gulls avoided the oil more successfully than the other species. However, a large number of the gulls had smudgy markings and the seriousness of this degree of oiling is unknown. The wintering population of birds in the bay was not expected to change until the spring migration began, possibly in about three or four more weeks. The concern was whether the beaches and rocky coast line would be substantially cleared of oil and whether the remaining oil in the submerged wreck would be removed without further spillage.

After a survey of oiled birds between February 13 and 16, effective searching became more difficult as snow covered the beaches and ice gathered in inlets and coves. However, small numbers of oiled birds continued to struggle ashore and dead ones to be washed up. Several oiled harbor and gray seals were seen. There were some reports of young gray seals being seen in the woods up to half a mile from shore. It was possible that these animals had become disoriented as a result of oil accumulations around the eyes and nostrils.

On February 27 the Department of Transport officials reported oil ashore on Sable Island, in the Atlantic about 120 miles southeast from the site of the wreck. A sample of oil was obtained and analyzed by chemists at the Bedford Institute of Oceanography. A measure of the vanadium content indicated that the oil was from the *Arrow*. From March 11 to 13 a considerable amount of oil was reported on the north beach and a lesser amount on the south beach. Along a total of 1.3 miles of searched shore line, the carcasses of 218 oiled birds were found. Most of these were alcids—murres, dovekies, black guillemots, and some common puffins. About 30 of the dead birds were fulmars, a pelagic, plankton-feeding species. It was the first report of this species having fallen victim to the spill. It was estimated that 5,000 birds had died and been washed up on Sable Island and that probably many more had perished but were not washed ashore as the slick was driven south past the island. Ten dead gray seals were found on the north beach.

By March 13 the many technical problems concerning the recovery of the oil from the wreck had been solved and pumping operations had begun. A spill of about 30,000 gallons occurred during the night of March 24/25. It was reported that this slick moved out to sea. To protect fish-processing plants and to prevent further contamination of the coast line, several containment procedures were adopted, including the strategic placement of booms of various kinds and the construction of a rock dam across Lenox Passage between Ile Madame and Cape Breton Island. In the cleanup operations, the task force resisted all suggestions to use chemical dispersants because of an inadequate knowledge of their total impact on the environment. Due to its high ignition temperature, most attempts to burn the oil, including the use of "sea beds," special flame-throwing equipment, and chemicals, failed. The low water temperature (34 degrees) and emulsification of the oil compounded the difficulties. Attempts were made to clean up sand beaches using heavy mechanized equipment. Peat moss, straw, planer shavings, and other materials were used with varying degrees of success to render slicks and onshore sludges more manageable. A conveyor belt device known as a "slick licker" performed well in the removal of oil from the water surface.

The Canadian Center for Inland Waters said that the *Arrow* incident generated a large amount of scientific research as well as a massive cleanup exercise.

Taking the Bitter with the Sweet: A Mixed Marriage

(Brazilian Bee Infestation: South America, 1956–73)

Beginning in 1956, *Apis* queens were brought from Africa to Rio Claro, São Paulo, Brazil, to develop a more productive, better-adapted race of bees than the European races then in use in South America.

In 1957, before the project was completed, 26 swarms headed by African queens escaped. Extensive crossing with the European bees obviously took place as certain attributes of the African race spread rapidly through the honeybees of Brazil and adjoining countries.

These Brazilian bees, hybrids of the African and European races, moved quickly into habitats where honeybees had previously been either scarce or absent and before long became the most prominent form of *Apis* in Brazil as well as in other parts of South America.

Good honey production and colony growth are their most likable characteristics while excessive aggression appears to be the most disliked. The "Brazilian bee" is dangerous, even fatal, to people and animals and exceedingly difficult to manage because of its unprovoked mass stinging and frequent swarming and absconding. Consequently, it has brought about drastic changes in the beekeeping industry in the areas it occupies in South America.

Although the Brazilian bees were found to show a great deal of variability, some general statements about their characteristics can be made. The positive and advantageous qualities they possess are observable in their flight of workers—bees that work in the cold and carry large loads; in their productivity, which appears to be high; and their development cycle, which produces rapidly increasing populations.

The characteristics which are considered to be disadvantageous include their behavior on examination of the hive; i.e., their behavior at this time is considered objectionable due to flying when the hives are opened, running on combs, and festooning. Additional negative characteristics concern excessive propolization; their seasonal life cycle, which displays poor winter-adaptive ability; and the killing of queens, upon disturbances such as jarring of the hive. Nest locations, because they have large feral populations forming in the natural nesting sites, are also considered disadvantageous, as is their tendency to excessive aggressivity in terms of stinging. Their keen ability to communicate alarm, necessitating a dispersal of hives, plus an excessive robbing characteristic sum up the objectionable side of the Brazilian bees. It is interesting to note that the last four characteristics are considered to be the major reason for their supplanting of the European races.

At present in South America, the Brazilian bee ranges from the equator to about 34 degrees south, indicating that it could well survive throughout Mexico and in the southern United States. Although the southward spread is now slow, the infestation is moving about 200 miles per year to the north and has crossed the Amazon.

From southeastern Peru to Macapá in Brazil, the bees are now 1,000 to 2,000 miles from Panama. At their present rate of speed, they will take five to ten years to reach Panama. However, if they are spreading from around Mount Roraima the distance is only 800 miles, and since they have already crossed the Amazon, progress may well speed up.

There is no known geological or climatic barrier that will prevent the spread of the Brazilian bee into North America. Because the southward spread of the bee is presumed to have been slowed by failure to winter well, it seems reasonable that it will not survive in the northern United States. Nevertheless, the entire continent would feel its influence, since queens and package bees are regularly shipped from the southern states.

Currently, the Brazilian bee has had an effect on people and public reaction, domestic animals, pollination activities, and consequent ecology of affected areas. Its effects have also been felt by the queen and package bee industry, honey production, and beekeeping in South America.

Potential means of preventing the spread of these bees into North America and of controlling it include the destruction of swarms and absconding colonies, quarantines, poison baits and sprays, bee diseases, and genetic barriers.

Snowmobilers Beware!

(Gray Fox Epizootic Canine Distemper; New York, U.S.A., 1972)

During the height of the winter season of 1972 in the Catskill region of New York State, snowmobilers suddenly found themselves the target of a new attack in the most literal sense. Calls flooded the regional offices of the state Environmental Conservation Department reporting "peculiar behavior" on the part of the gray fox with special mention of this animal's aggressive attention to snowmobiles and cars. The foxes had also been observed foaming at the mouth and attacking dogs, while just as often they appeared lethargic and totally oblivious to humans. Indeed, many dead fox were being found near human habitats.

Understandably, residents of the area began to fear the animals were rabid. However, the 20 animals received from Sullivan County by the Wildlife Research Laboratory were each diagnosed as having canine distemper. Further examination of all the heads by the New York State Health Department showed they were negative for rabies while the department also confirmed some of the animals as being distempered.

Although a number of deaths from canine distemper in gray fox and other carnivores had been observed in the southern Catskill region, there had not been many documented reports of epizootic canine distemper in some time—only isolated cases here and there.

Starting about February 18, 1972, in Sullivan County, cases of distemper began to appear in gray foxes and quite quickly seemed to spread to Orange County to the south and Delaware County to the north. By March 10 they were also showing up in Albany, Rensselaer, and Columbia counties with confirmed cases of the distemper found also in 5 raccoons, 2 skunks, and 2 red foxes.

From February through October 16, 1972, the Pathology Section at the Delmar Wildlife Research Laboratory diagnosed canine distemper in 164 foxes, raccoons, and skunks and 1 mink. As of November 21, 1972, cases were being positively diagnosed from all major regions of New York State.

These animals continued generally to exhibit the peculiar behavior originally cited: loss of fear of man, dogs, and dwellings; unco-ordinated movement and aggressive behavior; and in later stages, coma and convulsive seizures.

Winter is the breeding season of the gray fox and population was estimated to be high in February 1972 in the affected areas. However, as the epizootic went on, fewer and fewer tracks were observable in the snow and reports from trappers, fur buyers, and field technicians indicated highly decreased populations of gray fox, skunk, and raccoon following the distemper outbreak.

A final plaintive note to the report of this event stated that populations of rarer carnivores such as the fishers may have been seriously affected by the disease though there was little likelihood of locating specimens.

More Grass per Day Than a Cow

(Paraguayan Leaf-cutting Ant Plague: South America, 1972)

Forty families were recently ousted from their homes and forced to abandon their property in the village of Filadelfia in central Paraguay because of damage by leaf-cutting ants.

Certain areas of the South American Gran Chaco are showing high infestations (up to 5 nests per hectare) of *Atta vollenweideri,* leaf-cutting ants endemic in the Chaco.

These ants forage mainly on grass species and ten nests of them can consume more grass per day than one cow. Obviously, this adds up to a considerable loss of grazing pastures, especially the most valuable ones.

In consequence, the government of Paraguay has decided to open up this relatively isolated territory by paving the dirt road and building a bridge over the Paraguay River—an operation to be completed by 1977.

The next steps to be taken have been defined in terms of a broader frame of reference. They are: a search for a less expensive poisonous bait than the one presently used; a study to determine the economic importance of the damage by the ants on cattle raising; and the effect of the ants on the total ecology of the infested areas.

A Jaundiced View

(Miami Palms Lethal Yellowing: Florida, U.S.A., 1971, 1972, 1973)

A woman from the Palm Society of Miami reported that several palms were losing their coconuts and subsequently dying in the Coral Gables area of Miami in September 1971.

A plant pathologist from the Florida Department of Agriculture was immediately dispatched to examine the trees. A check of symptoms revealed a perfect match to a disease known as "lethal yellowing" which was first described in Jamaica in 1890. Since then it is known to have moved into other parts of the world, including central South America and Africa, and was reported in Key West in 1955.

As of January 19, 1973, the disease was restricted to Miami proper and the Dade County area; i.e., the occurrence was scattered across Miami into Miami Beach and the Key Biscayne area just east of the Miami mainland. Thus the spread from the original infestation at Coral Gables was approximately twelve miles to the north, four miles to the east, about the same distance to the south, and two to three miles to the west.

Of an estimated 400,000 palms in the Dade County area, the total of infected trees was calculated in October 1972 to be approximately 1,500, based on monthly surveys. There are about 1 million coconut trees on the Florida mainland—all potential victims if the disease is not quickly brought under control.

A plantation program was instituted in which diseased trees were cut down and burned as quickly as they were observed to be displaying symptoms. This program is currently in operation despite various prior halts due to funding difficulties.

One of the more interesting aspects of lethal yellowing is that the means of introduction of this disease is based on pure speculation. At this writing, the causal agent has never been identified although recent testing and the process of elimination tend to indicate either a virus or a microplasma-like organism. How it is transferred from tree to tree is not known either. However, it is theorized that it is probably by an insect or a mite. It seems, too, that the coconut palm is the only species to be affected.

As previously mentioned, the first indication of the presence of the disease is when the plants begin to lose their coconuts. Then the newly opened tip of the flower becomes black and shriveled and the leaves begin to yellow, usually from the lower fronds up, until eventually all of the fronds turn yellow and die. Finally, parts of the tree fall off, leaving only the trunk.

All of the above occurs within a period of one to three months from the time the beginning symptoms are first seen. Consequently, with its ability to kill a tree in such short order, lethal yellowing appears to be a dangerously fast-working and devastating disease. Once a tree is hit, it cannot be saved. Everywhere the disease has struck, it has wiped out huge numbers of palms. In Key West (1955–68), out of a total of about 20,000 palms, the epidemic killed 75 per cent or 15,000. It also wiped out many whole coconut plantations in Jamaica, where, incidentally, resistant varieties have been replanted on many of the plantations, bringing them back into production.

At present the Institute of Food and Agricultural Sciences at the University of Florida and the Florida Department of Agriculture's Division of Plant Industry are co-operating to see what can be done about the disease. The services of the U. S. Navy in Key West were also made available and have been utilized during testing there and in connection with the surrounding community during the epidemic.

Since the disease undoubtedly poses a serious threat in the state of Florida, immediate action is being taken to obtain data sufficient to determine the rate of spread. In an effort toward control, discussion is also under way concerning possible replanting with the resistant Malayan dwarf variety of coconut palms.

On January 29, 1973, a report from the Agricultural Research Center at the University of Florida revealed a total of 3,168 cases of lethal yellowing diagnosed in Dade County at the end of December 1972 with the apparent rate of infection calculated at 0.2 per cent per month. The report also contained the following ominous forecast: "If the epidemic continues at this rate, 99% of the coconut palms in Dade County will be dead by early 1977."

Somewhat more encouraging was the news that, since therapy with tetracycline antibiotics has given symptom remission in other plant diseases thought to be caused by mycoplasma-like organisms, this approach was tried with lethal yellowing. Note: Mycoplasma-like bodies have been reported in the phloem tissues of lethal-yellowing-affected coconut palms in Jamaica by three separate sources in the past year.

It was found that symptom development could be repressed for periods of five months or more in trees showing initial symptoms of the disease. Without treatment, such trees would normally be dead within four months.

"The possibilities of using this treatment on a large scale are presently being explored."

Like the Waters of Ancient Egypt

(Massachusetts Red Tide: Massachusetts, Maine, New Hampshire, U.S.A., 1972)

A reddish-brown mass was noted in the water off the Massachusetts North Shore (from Cape Ann north) on Thursday morning, September 14, 1972, by a biologist from the University of Massachusetts Marine Laboratory in Gloucester, Massachusetts. A sample of this water was taken and an extract distilled from it and injected into two mice, which were dead within eight minutes. An extract was then taken from clams and mussels and subsequently injected into four other mice, which also died. A "red tide" was immediately reported.

A species of algae causing red tides in the cold water off Canada is known, as is another species in the southern warm waters of Florida and California which kills fish. However, the appearance of a red tide in Massachusetts is a highly unusual event since Massachusetts is ordinarily too far south for the Canadian species and too far north for the southern variety.

Nevertheless, the organisms in the water off the North Shore were identified as the Canadian species *Gonyaulax tamarensis,* a dinoflagellate form of planktonic algae that is taken in by clams and other shellfish. It is not toxic to the shellfish, but it accumulates in their digestive tracts where it becomes concentrated. The toxin produced by the algae is, however, poisonous to warm-blooded organisms, in which it causes paralytic shellfish poisoning. Early reports confirmed twelve cases of this in humans in the Massachusetts area as of September 18, 1972, and a number of sea gulls and ducks were found to have died of this toxin.

Some shellfish were affected in the shore area, but this was believed to be due to other factors that occurred simultaneously with the red tide, possibly the same factors that promoted the algae bloom.

According to the U. S. Fish and Wildlife Service, by September 18, about 175 dead black ducks (mostly adult males) were found in the area affected by the red tide. Six birds on Tuesday (9/12); on

Wednesday (9/13), it had jumped to 50; by Thursday night (9/14), the count had reached 118; and on Saturday (9/16), it was 157. A few more birds were found Sunday (9/17) and Monday (9/18), bringing the total to 175. Approximately 87 gulls were also found in addition to some incidental birds, including 2 black-bellied plover, 1 black-crowned night heron, and 1 green heron.

When the dead birds were tested for hard pesticides and these were definitely ruled out, the bird deaths were thought to have been caused by the red tide organism.

No mammals were reported as having been affected. Indeed, there appears to be no reason why they wouldn't be. But although musk-rats, mink, or otter might feed on shellfish, none turned up as victims of the red tide. The total list of affected organisms did include a few crabs, clams, and the birds.

On Friday, September 15, the red tide was reported stretching from Manchester, Massachusetts, north to Portland, Maine. It was .5 to 3 miles offshore and formed a 12½-mile-wide band. That length of coast line would be roughly 75 miles.

At this time it was thought that the plankton bloom might have been caused by recent unusual climatic conditions. The organism (*Gonyaulax tamarensis*) exists in the affected area in varying densities at all times, but normally does not exist in densities great enough to cause any serious problems.

On September 20 it was reported that the Department of Public Health had closed all clam flats in Massachusetts for digging shortly after the red tide was discovered. Restrictions were also put on other shellfish as a safety precaution, but unaffected species of shellfish were gradually released for sale, e.g., hard-shelled clams such as quahogs, cherrystones, and littlenecks as well as Cape scallops and oysters. The species on which a moratorium was never declared included fin fish, lobsters, shrimp, and crabs. It was primarily the soft-shelled clams and mussels that were affected and these were the two for which there was no digging, no sale, and no export or import.

These regulations went into effect progressively, and by the twentieth the federal Food and Drug Administration was involved in the situation, especially with regard to recalling shipments that might have gone out.

According to the Department of Public Health, by Monday night, September 18, there were 25 confirmed cases of paralytic shellfish poisoning (PSP) in Massachusetts. However, at that time no fatalities were reported. The public now began to hear something about the symptoms of PSP, which, though rarely fatal in humans, include numbness and tingling, with vomiting likely in more severe cases and respiratory difficulties occurring only in the most severe attacks.

The only place where sampling had verified the presence of the red tide organism was in the area of the initial infestation in Massachusetts on the North Shore from just south of Cape Ann north into New Hampshire and along the coast of Maine. Although there was speculation on the twentieth that shellfish in other areas might be accumulating the toxin produced by the algae, only a few limited samples were taken in which some of the toxic organisms were found, so that this theory was not adequately verified.

Water samples taken by the University of Massachusetts Marine Laboratory scientists from the North Shore on Sunday, the seventeenth, showed negative evidence of the red tide organism in that area and indicated a dissipation, though the affected shellfish were expected to remain toxic for some time—ten days to three and a half weeks for soft-shelled clams and possibly all year round for some other animals.

It was also reported on the twentieth that not much water sampling was taking place in Massachusetts at the state level, the state being primarily concerned with verifying the extent of the contamination at that time. Consequently, an intensive sampling of shellfish (especially soft-shelled clams and mussels) for the toxin was being conducted and overseen by the Department of Public Health throughout the state south of Cape Ann, north of Cape Cod, on the North Shore, etc., with a continual monitoring in suspected areas to determine the rate of decrease of toxicity.

On the morning of September 21 it was reported that there were still streaks of the red tide out approximately 6.5 miles. The total bird mortality now included an additional 96 black ducks, 37 herring gulls, and 1 greater scaup, found mainly in the Merrimack estuary.

At this time the Marine Fisheries Division of the Massachusetts National Resources Department reported that the red tide seemed to have dissipated over all of the originally infected areas since the counts for the organism in water samples had returned to the normal level. It appeared that the red tide had peaked on the fourteenth or fifteenth and the count had been going down ever since. Nevertheless, the shellfish were still contaminated and it was not known how long this condition would last.

In Massachusetts the red tide was confined to the North Shore from Cape Ann to the New Hampshire border. One other bad sample was obtained from Chatham, Massachusetts, on the nineteenth but all along the entire South Shore to Cape Cod and from Chatham north to Cape Ann no other red tide shellfish contamination had been found. The outbreak which occurred at the mouth of Pleasant Bay was apparently only a local occurrence caused by favorable environmental conditions in that area. It was not believed to be a spread down from the north.

In connection with the ban on shellfish, digging was still prohibited throughout the state as of the twenty-first but some sale of shellfish was being permitted in Massachusetts. Retailers who had shellfish on hand were allowed now to sell them. The sale ban was lifted on quahogs on the twentieth and on soft-shells and mussels on the twenty-first.

The Department of Sea and Shore Fisheries in Augusta, Maine, reported on the twenty-second that they had first become aware of the present problem on September 14. They then began to monitor the area on the western part of the coast. Heavy concentrations were found in the western part of the state, with sporadic but high counts also in other parts of the state, thus displaying an inconsistent, unpredictable distribution. Monitoring was being continued to determine a return to safe non-toxic levels in the affected clams, quahogs, and mussels. However, on the twenty-first counts were still high. Note: It is necessary to close off areas when the counts reach 80 and they were receiving counts up to 3,000 to 4,000.

This event was described as an unusual circumstance for the state of Maine, since this problem had never previously been reported in the western part of the state (Cape Elizabeth and west) during the previous twenty years of the department's existence.

The Deparment of Sea and Shore Fisheries regularly monitors eastern Maine because periodically and not uncommonly toxic algal problems are to be found there. Generally, it has been down in the Casco Bay area with an occasional western extension. In 1961 it extended as far west as Pemaquid Point on the central part of the coast. Parts of eastern Maine were indeed closed for about three weeks during August 1972 and a couple of areas on the coast are kept closed all the time because of sporadic and unpredictable outbreaks of the problem. These are, however, very small areas involving offshore islands.

The department reported that in the past twenty years in eastern

Maine they have never found toxic levels associated with a visible red tide. The toxic problems for which they have closed in the past were invisible—the reason for their very frequent monitoring routine. In addition, they have had red tides on other parts of the coast which were not toxic, so they were not at all sure that what they were seeing was actually causing the problem. They felt they might be parallel developments in the water but unrelated developments as far as toxicity was concerned. They maintained they did not know what was causing the red color and were awaiting a report from the University of Maine on the water samples. "*Gonyaulax tamarensis,* as far as we know, is the organism in the clams; it has been in eastern Maine in the past. The people in Massachusetts tell us that it is *Gonyaulax* and that it is a pigmented organism."

At this point, in Maine, the Department of Sea and Shore Fisheries (marine aspects) and the Department of Inland Fisheries and Game (birds) were the two agencies directly involved. Other agencies in Maine indirectly involved were the Department of Public Health and Welfare (disposal of mollusks), Department of Agriculture (selling the affected mollusks at the retail level), the Environmental Protection Agency (disposal of clams), and the Governor's Office (relief for clam diggers).

The Maine Department of Sea and Shore Fisheries concluded its report of the twenty-second with the encouraging statement, "Canada does not have a red tide problem as far as we can determine."

On the twenty-second, the Maine Inland Department of Fish and Game reported that no damage to game of any species (including birds) had been detected in Maine, though they were seriously concerned with monitoring the problem in connection with sea ducks, particularly the eiders and scoters. Their scoter and eider distribution at that time was primarily north of the major sea duck areas. The red tide there was primarily offshore in the worst places—Matinicus Island being one of the major areas where it was particularly bad, and that about twenty-five miles offshore.

The Federal Bureau of Sport Fisheries and Wildlife report of black duck and herring gull mortality in Massachusetts evoked the following information. "So far, all of the dead ducks in Massachusetts have been adults. Infant black ducks prefer vegetable food but during their first hard winter, they learn to eat clams, mussels and marine snails. The adults, learning to prefer the mollusks, continue to feed on them thereafter. Secondary poisoning occurs in the herring gulls that feed on the poisoned ducks when their intestinal tract is eaten."

The New Hampshire Department of Fish and Game reported on the twenty-fifth that closure of the New Hampshire clam flats and oyster beds on the advice of the New Hampshire Health Department had been in effect since September 15 and remained so. Shellfish were still being tested for the toxin and it was expected this would continue. At that time, some 6 cases of illness as a result of eating clams had been reported.

The occurrence of the red tide had first been observed offshore in New Hampshire on Thursday afternoon, September 14. Although red tides do occur annually in New Hampshire, they are usually far offshore and do not cause any damage. This time, however, the entire New Hampshire coast was affected.

Intensive testing was continuing at the University of New Hampshire in Durham by a scientist who had prior experience with red tides on the West Coast and the Gulf of Mexico.

The New Hampshire Department of Fish and Game report also observed that some gulls, black ducks, and other birds had died as a result of the red tide and these specimens had been sent on to various federal laboratories for testing.

On the twenty-fifth, the Massachusetts North Shore was still closed and it was reported that the toxicity level of the shellfish had not dropped to any significant degree.

The red tide had apparently disappeared as far as aerial observation could tell. It seemed to have moved offshore and broken up. Certain selected key stations along the Massachusetts coast were being used for a surveillance network to check shellfish for some time. The stations on the North Shore were in Gloucester, Essex, Ipswich, and Rowley. Pleasant Bay was also being monitored though the level there had not been high.

The Massachusetts Department of Public Health in consultation with the Department of Natural Resources announced that, beginning on the twenty-fifth, commercial harvesting limited to quahogs, or hard-shelled clams, would be permitted in selected areas in southeastern Massachusetts. As of September 24, the North Shore bird mortality total had risen to 328 black ducks and 152 herring gulls, according to the Parker River National Wildlife Refuge.

The Maine Department of Sea and Shore Fisheries reported on the twenty-ninth that, as of the thirtieth, the Canadian border would be open. The remaining part of the coast from Cape Elizabeth to the New Hampshire line would remain closed and continue to be monitored.

The mollusks in Maine that showed particularly high toxicity scores were soft-shelled clams, surf clams, and the non-edible portions of the sea scallops.

Their toxicity levels started to drop about September 25 or 26, and on the twenty-ninth they were reported back to normal except for the offshore islands previously mentioned and the western portion of the coast. In the western area, the toxicity level had dropped about 50 per cent in some locations while others were still high. However, even in areas where there was a drop the levels were high enough to keep the areas closed—the counts were running about 1,000 micrograms with 80 micrograms being considered the cutoff point.

At the time of this report, the water had not been checked for the red tide during the last few days but on the last aerial flight there were some places where it still showed up. Some interesting conditions were noticed to be in bloom at this point. In eastern Canada and Maine, which have been the historic centers for the outbreak, the organism *Gonyaulax* was not visible in the water. According to the Canadians, it is, however, visible as a brownish tan in almost pure cultures.

The Institute of Natural Environmental Resources (at the University of New Hampshire) report of October 2 listed the red tide mortality of 20 black ducks, 10 herring gulls, and 1 shore bird as of September 27, after which time no other birds were brought in.

On October 2 the water in New Hampshire had the typical phytoplankton. The level of *Gonyaulax* had dropped back, probably since September 24, and was remaining relatively normal. The soft-shelled clams were still toxic though the level had gone down.

The clams were being monitored and the water checked daily. The toxicity level peaked in that state on September 17 and 18, and as of the twenty-ninth had dropped to about one half of what it was.

The Department of Zoology at the university reported on October 2 that they were serving as the bio-assay lab for the state of New Hampshire and northern Massachusetts points. Standard bio assays were run and the clam flats checked daily. Helicopters were used by the Coast Guard and Fish and Wildlife people to check for the presence of the red tide. Bird meat, oysters, and other organisms as well as clams were checked to determine their toxicity levels. The oysters tested were from Great Bay. When tested on the seventeenth, they were non-toxic. However, when tested again on the twenty-first (also from Great Bay), they were slightly toxic, though definitely over the safety level.

They observed that this was the first time New Hampshire had been the site of the *Gonyaulax* phenomenon. Before this, some New Hampshire botanists had made transects out to the shoals but only occasionally picked up a couple of cells here and there.

During the red tide, the cell counts off Hampton were between 2,000 and 10,000 cells/ml water (2–10 billion/1) until about the sixteenth and into the seventeenth of September. Then it began to drop down to 1,000 cells/ml (1 million/1) and then further. The cell density varied quite a bit along the coast. "There would be a streak that went for probably a mile or two and then broke with another streak not far away. Some of the red tide came within 100 yards of the shore."

The Massachusetts Natural Resources Department reported on October 3 that since the September 23–24 weekend there had been no new illnesses and the Department of Health was then beginning to restore some of the fisheries. The commercial harvesting of quahogs was allowed in all areas except Pleasant Bay, Stage Harbor, and the Nauset Inlet complex—all on Cape Cod. Some shellfish (not quahogs) evidently had levels of contamination though thus far no scallops had been contaminated in any of the areas sampled. There had been some contaminated viscera in the Pleasant Bay area.

On September 30 the Department of Public Health allowed the opening of all family and commercial shellfishing in all areas except the North Shore; the same areas that were open for commercial harvesting of quahogs were opened to the harvesting of all shellfish except soft-shelled clams and mussels with the exception of the area mentioned.

On the North Shore, the department reported the decrease of toxicity levels in soft-shelled clams and mussels. There was no evidence of any further outbreaks of the *Gonyaulax* organism so it was anticipated that this area would be opened within the period of a few weeks.

Sunday, October 1, was the opening of the bay scallop season but it was recommended by the Division of Sea Fisheries of the Natural Resources Department that the season be postponed for two weeks for two reasons: first, for fear of a depressed market and, second, because of the quality of the scallops themselves—when there is a warm fall, the adductor muscle seems to be soft and not as large.

Finally, in its October 3 report, the department stated that it felt the red tide in Massachusetts was initiated in the area of Gloucester, Rockport, and Newburyport and that there was the possibility of a northward progression. However, with regard to the instance on Cape Cod, it was felt this was an isolated outbreak with conditions right for a great increase in the organism. Other areas were sampled, and although a slightly toxic mussel was found in Scituate, there were no problems with any clams there.

Apparently one of the things that presented a problem to the department was that there had been no sampling for the organism in the past, so there was no basis for comparison. However, it is known that certain conditions must prevail for increases of such high proportions in the *Gonyaulax* organism. These conditions include lowered salinities and a higher water temperature. The organisms do seem to be able to bloom in a low nutrient situation, so nutrients may or may not be important if high.

The October 3 report of the Parker River National Wildlife Refuge in Massachusetts provided the last bird count for the Massachusetts North Shore (to September 27). It included the following totals: 364 black ducks, 175 herring gulls, 1 eider duck, 3 white-winged scoters, 1 American scoter, 1 mallard, 2 black-bellied plover, 1 black-crowned night heron, and 1 green heron.

Part III
ASTROPHYSICAL EVENTS

A Blue Disk the Size of the Moon

(The Tunis Fireball of 1969 and Other More or Less Genuine Visitations from Space)

In 1969 a number of reports were received from Tunis concerning the appearance in the night sky over North Africa of a blue disk the size of the moon. During a thirteen-minute period on July 6 the disk was seen to change its shape slowly, passing from a circular to an elliptical outline and finally to that of an expanding cloud. It grew, blurred, and blended with the darkness of the sky.

Another object suddenly appeared in the north about 10 degrees east of Polaris. It was as large and bright as the evening star. Soon after, it distended and burst like a nuclear fireball. Around a luminous center the size of the moon, there formed a glowing shock wave, blue in color, which grew rapidly outward and at the end reached the diameter of a very large lunar halo. In a fraction of a second, the luminous ball gave way to a misshapen cloud which faded slowly. This second phenomenon lasted only half a minute. No noise was heard throughout.

As it later turned out, the "fireball" was a glowing cloud which resulted from atmospheric testing by the Europeene de Recherche Spatiale. It was caused by the launching of Skylark rockets from a base in Sardinia on July 6.

It is understandable why the observers thought they were seeing a meteoric phenomenon. Many of the antics of the atmospheric rockets that night resembled the real thing. By expanding their laboratories into the skies, scientists occasionally succeed in confusing even themselves, if only briefly, between their own experimentation and a bona fide happening from outer space. The effect on casual observers may be even more puzzling and of longer duration.

Some similar instances of man unwittingly imitating astrophysical phenomena occurred over South America, the United States, and Europe in approximately the same period as the Tunis Fireball:

The Comodoro Rivadavia Fireball (Argentina) of August 17, 1968, was later identified as the re-entry of Satellite 68-2B, Cosmos 211, launched on April 9 of that year.

The North Haledon Fireball (New Jersey) of August 2, 1970, was subsequently determined to be an aerial balloon with a flare or some type of slow-burning device that emitted sparks.

The Ludwigshafen Fireball (West Germany) of February 21, 1971, was described by many witnesses as being typical of a meteorite fall at low altitudes. Two days later a French newspaper published an article on a rocket experiment that might have accounted for the phenomenon, although the connection between the two was never clearly established.

Other fireballs are also in the suspect category as man-made debris from experiments many months or years old. There is no doubt, however, about the credentials of the Mount Riffler Fireball that entered the earth's atmosphere above Germany and Austria on November 24, 1970, and flew a luminous trajectory of 108 miles in 9.5 seconds. It was photographed successfully by eleven German stations and one Czech station of the European all-sky camera network and tentatively classified as a stony meteorite with a probable linear dimension of 4 inches. A search of the general impact area was conducted by the Max Planck Institut für Kernphysik without result.

Comets and meteors have been noted from the earliest times, by the Chinese as many as twenty-six hundred years ago and by Europeans for fifteen hundred years or more. It was believed that, like all heavenly bodies, they influenced the affairs of men and heralded plagues, famines, wars, and other things invented by humans to destroy themselves. The early records, therefore, were kept out of fear and superstition and were of interest to the enlightened scientists when they finally systematized observations and study. Great meteor showers were regarded with special dread and foreboding, and there was one of particular brilliance in 1833 that could be viewed from all parts of North America. An estimated 200,000 shooting stars were seen from one vantage point between midnight and dawn. Many left persistent trails, yet there is no evidence that any reached the earth. At the time there were a number of fundamentalist religious sects in existence, such as the Millerites, which believed in the imminence of the Second Coming. The effect of this spectacle on them can well be imagined.

Science has not yet graduated to a full understanding of every celestial apparition. On June 9, 1970, two scientists of the Uttar Pradesh State University, India, were making telescopic night observations at the observatory near Naini Tal, about a hundred miles east and north of Delhi. Just before 8:30 P.M., Greenwich Mean Time, they noticed a sudden brightening of the sky lasting about one second. The night was cloudless and moonless, with a relative humidity of 46 per cent, a temperature reading of 16 degrees C., and a wind velocity of almost zero.

The two men reported that the illumination reached the brightness of lightning in two stages of increasing intensity and appeared to originate in the west or southwest. It involved the entire local sky, raising it to light intensity comparable with that of the Milky Way during the summer months. In fact, the Milky Way merged within this light for the period of its duration.

Neither of the observers could find anything to which this phenomenon could be related. The Naini Tal Sudden Sky Brightening is still a mystery, although devoid of the frightening implications it would have had in past centuries.

Meteors, for their part in the astrophysical side show, have yielded up a good many of their secrets since they became the objects of serious scientific investigation about a hundred and fifty years ago. In their fall, they are delightful and, in the main, harmless visitors to earth. According to published figures, only 10 meteorites are known to have struck buildings between 1847 and 1960. Between 1968 and 1971, however, the record book was livened up with 5 more documented instances. The locations are diverse: New York State, Connecticut, Czechoslovakia, Finland, and Australia. Farmhouses and barns were singled out in the last three places for some unexplained galactic reason. No one was hurt, although there were some close ones.

Meteorites (or meteors which outlast their fiery plunge to fall to earth as a solid body) do not strike with an insane cosmic force that disintegrates everything in their path. The speed of falling meteorites is considerably slowed by the earth's atmosphere to about that of a stone dropped from the same height—400 to 600 feet per second. Although standing directly beneath is not recommended, the impact at this velocity is not necessarily catastrophic. A mass of iron which fell on a house in Bohemia in 1847 penetrated the roof and showered debris over a bed in which three children were sleeping.

A 645-pound stone which fell in Bohemia in 1866 (meteorites *always* seem to be falling on Bohemia) made a hole in the ground 11 feet deep, impressive but not unreasonable in view of its weight. At the smaller end of the scale, a meteorite weighing about 5 kilograms and measuring 12 by 6 inches in linear dimensions, made no discernible impact crater when it fell in a field in Venezuela. The farmer who picked it up immediately afterward found it only warm to the touch.

There are, of course, pyrotechnics associated with the larger packages. A 20-ten meteorite fell near Blackstone, Virginia, in 1922 with an appalling explosion and tongues of flame. From this can be inferred something of what must have happened if the craterlike depression in Arizona known as Coon Butte was actually created by a meteorite, as some maintain it was. Coon Butte measures 4,000 feet across, is 550 feet deep, and has walls of limestone rising 100 feet above the plain.

Most meteorites, however, are miniature in relation to their apparent size in flight and the sonic disturbances they make. Even the brightest are often little more than a fraction of an inch in diameter and a few milligrams in weight. They all appear larger in their trajectories, owing to the envelope of heated and glowing gas that surrounds them. It is not unusual for a meteoroid (a meteor that enters the earth's atmosphere) to attain the apparent diameter of the moon and to shed as much light. Some are reported to make the night as bright as the day, and although this might be an exaggeration, it is a pardonable one in view of the contrast with the darkness of the skies.

About 20 million meteors are estimated to enter the earth's atmosphere daily. The overwhelming majority burn themselves out before reaching the earth's surface. The most spectacular of these are known as fireballs or bolides: because of their greater mass they are able to penetrate lower before destruction. When they overtake the earth before midnight, following its rotational path, they sometimes have trajectories many hundreds of miles long and a relatively long survival span. Their passage is frequently accompanied by sonic booms, as many as three or four in succession, and those with extremely long paths may pass from horizon to horizon, seen by many observers and, in modern times, photographed by many camera-tracking stations in turn. The end of the fireball is frequently marked by a huge detonation as it splits into innumerable smaller fragments strewn over a wide area.

With this modest background drawn from the vast technical lore which exists on meteors and their offspring, it is possible to gain some perspective on a choice few of the more recent visitations from space.

A Trail of Sparks with a Buzzing Sound

(The Veracruz Fireball: Mexico, 1968)

Three huge meteorites turned the night sky into day and shook Veracruz when they crashed into nearby hills at 2:15 A.M., local time, March 27, 1968. A large number of observers had seen an intense light changing from blue to red and making a dazzling traverse of the sky.

The meteor appeared as a brilliant globe with a trail of sparks behind, accompanied by a buzzing sound.

Crews of fishing vessels nearby said the meteorites began as greenish stars which turned the night sky a brilliant white, and then became blue just before impact.

A Harvest of Meteorites

(The Allende Meteorite Shower: Mexico, 1969)

Between 8:05 and 8:10 A.M., Greenwich Mean Time, on February 8, 1969, a blinding blue-white fireball turned night into day over a 1,000-mile path from southern Mexico to El Paso, Texas. Hundreds of observers throughout Mexico reported seeing the brilliant flash of light and hearing a tremendous explosion. The most spectacular phenomena were centered around the city of Hidalgo del Parral in the south-central part of the Mexican state of Chihuahua. The fireball approached from the south-southwest and was accompanied by detonations and a strong air blast. Air Force meteorologists calculated the wind direction and velocities at the time of the fall, and a B-57 was flown through the probable dust train, which by then had drifted over the Gulf of Mexico. Samples of atmospheric dust were collected by special filter traps aboard the plane.

Thousands of meteoritic stones rained down over a large area of rural Mexico. Individual stones were recovered and identified almost immediately, but the extent of the shower was not realized until much later. Within a few weeks after the fall, the Smithsonian Institution's Division of Meteorites, in Washington, D.C., distributed Allende material to 37 laboratories in thirteen countries. Other organizations also distributed material.

The strewnfield lies largely to the south of the Interamerican Highway (Route 45) halfway between the cities of Parral and Jiménez. Its axis extends 31 miles (27 degrees N. and 105 degrees 12 minutes W.), and its area may be greater than 185 square miles.

The largest specimen found during early exploration was at the extreme northeast end of the field, 110 kilograms (about 242.44 pounds) broken into many pieces on impact. Specimens generally decrease in size toward the southwest within the field.

This strewnfield appeared to be the largest on record and the most productive of meteoritic material. An estimated two tons was removed from the area initially, and conservative estimates were that the Allende meteorite, a Type III carbonaceous chondrite, reached the ground with at least four tons of material.

To Catch a Falling Star

(The Ozarks Fireball: Oklahoma, Kansas, Missouri, U.S.A., 1969)

One of the most spectacular and, at the same time, elusive meteoroids ever to play tag with a battery of tracking cameras is remembered by the name of the Ozarks Fireball.

At 7:17 A.M., on October 9, 1969, an exceptionally bright meteor appeared over Oklahoma, Kansas, and Missouri after moving around the sun in a strongly elliptical orbit. It was photographed by no fewer than nine of the sixteen camera stations of the Prairie Network, operated by the Smithsonian Astrophysical Observatory. This network was established primarily to determine the geographical locations of possible meteorite falls from fireballs.

This fireball was so brilliant that one photograph taken at Pleasanton, Kansas, at a distance of 35 miles, shows some image reversal. Nearly 400 miles away, at Urbana, Illinois, a large orthicon responded to the meteor's light, even though it must have been on, or below, the horizon. The orthicon, a TV-like tube on patrol for meteors as faint as the eighth magnitude (the brightest are first and second magnitude), was pointed at about the center of the arc of the fireball, but at an altitude of 45 degrees.

Initially, the Ozarks Fireball must have had a mass of tons. Penetrating the atmosphere to within 15 miles of the ground, it caused sonic booms over a wide area. It then exploded and, in the last tenth of a second, disappeared completely, virtual proof that no sizable fragments survived the terminal bursts.

The Farmer Was Searching for a Sick Calf

(The Lost City Meteorite Fall: Oklahoma, 1970)

A fireball brighter than the full moon descended over northeastern Oklahoma at 8:14 P.M., Central Standard Time, on January 3, 1970. It was seen from as far away as central Nebraska. Traveling east southeast, it caused sonic booms that were heard from Tulsa to Tahlequah, sixty miles apart. The fireball dropped a 9.85-kilogram (about 22 pounds) meteorite that was found on January 9.

The Lost City Meteorite is the first to be recovered in a search that was guided by the photographically determined trajectory of a meteor through the earth's atmosphere. It was only the second meteorite whose orbit around the sun, prior to encountering the earth, was known from photographic observations.

The Oklahoma fireball was moving at about 9 miles per second when it entered the earth's atmosphere. It had decelerated to a little more than 6 miles per second when it was 18.5 miles high, and to 2 miles per second when it ceased to be visible at an altitude of 11 miles. The latter two are record low speeds for meteors. Moreover, the photographs do not show the flaring common in bodies which are disintegrating in flight.

Highlighting the advances that have been made in tracking meteorites are two calculations which were based on observation:

Assuming that the mass had a density of 3.4 grams per cubic centimeter, calculations of the impact point were made. It was predicted that, if the body were a sphere roughly 1 kilogram in weight, it would have reached the earth about three miles east of Lost City, Oklahoma. This tiny farming community is in Cherokee County, about forty-five miles east of the city of Tulsa.

Starting just eighteen hours after the event, a plane flew at 60,000 feet from Oklahoma City downwind to Atlanta, Georgia, and back, collecting upper-air samples to recover meteoritic particles. This attempt met with considerable success, and a large number of ablation particles were obtained.

The first specimen was found near Lost City, as predicted, and flown to the laboratories of the Smithsonian Institution in Washington, D.C. There, the 22-pound fragment was identified as a bronzite chondrite. It was an oriented stone, meaning that the flow patterns in the fusion crust indicated that the meteoroid did not tumble during the last portion of its descent.

The mass-to-area ratio of the recovered sample was 26 grams per square centimeter. A value of 20 had been deduced from the fireball photographs.

On January 17 a farmer searching for a sick calf found a second fragment, 10 ounces in weight, half buried in the sod. Smithsonian Astrophysical Observatory investigators found two more fragments, as they had the first, and the total mass recovered was 17.3 kilograms.

The field investigation began almost immediately after the probable impact point had been determined from the photographs of the fireball. Because of the excellent trajectory data and the recovery of the airborne particles, the Lost City fall represented an excellent opportunity for studying ablation at high velocities.

It Shook the Houses and the Cars

(The Bovedy Fireball: Northern Ireland, 1969)

On the evening of Friday, April 25, 1969, a meteor was widely reported to have passed over southern England, Wales, and Northern Ireland. The sound effects it produced were heard by many people over a wide area, and most observers agree that the approximate time of passage was between 9:20 and 9:25 P.M.

The fireball, caused by the meteor's passage through the atmosphere, was first seen over southeast England and North Wales. It passed overhead at Cardiff and Aberystwyth at an estimated height of about 40 miles. Continuing northwest, it passed to the west of Anglesey at a height of about 32 miles; it next crossed the South Down coast of Northern Ireland, approximately midway between Ardglass and Newcastle, and then over Tyrella Beach in Dundrum Bay.

Observations up to this point gave the basis for calculations which put the over-all diameter of the fireball at 660 feet. Nearly all observers reported a tail between 5 and 10 degrees in length, and the presence of a number of smaller bodies, average 3, following the nucleus at about ¼-degree intervals.

Having crossed Dundrum Bay, the fireball continued northwest, passing almost directly overhead at Ballynahinch, and about six tenths of a mile to the west of the town of Lisburn where, at Sprucefield, a fragment weighing a little under 1 lb. reached the ground. It penetrated the asbestos roof of a shed at a Royal Ulster Constabulary supplies depot. Part of the roof fused into the crust of this fragment. The main body went on its way a few miles to the west of Antrim and overhead at Kilrea, where the largest fragment was recovered in a meadow at Bovedy. This weighed about 12 pounds and was found in a pit just over one foot across and under 10 inches deep. The vertical depth of penetration was 1 foot, 2.5 inches, at an angle between 30 and 50 degrees from the horizontal.

The sonic boom and shock wave was both heard and felt in many parts of the country and was so intense in some locations as to shake

13. The carcass of a red-breasted merganser, among the thousands of sea birds mired in an estimated 1.5 million gallons of oil which, by February 14, 1970, had spilled into Chedabucto Bay, Nova Scotia, from the grounded tanker *Arrow. Photo: T. J. Wood, Wildlife Biologist, Canadian Wildlife Service, Fredericton, New Brunswick, Canada.*

14. A stricken raccoon, its eyes streaming in a characteristic symptom, is among the victims of the epizootic canine distemper that was fatal to the gray fox and other members of the animal kingdom in New York's Catskill region during 1972. *Photo: J. Corbett, New York State Department of Environmental Conservation.*

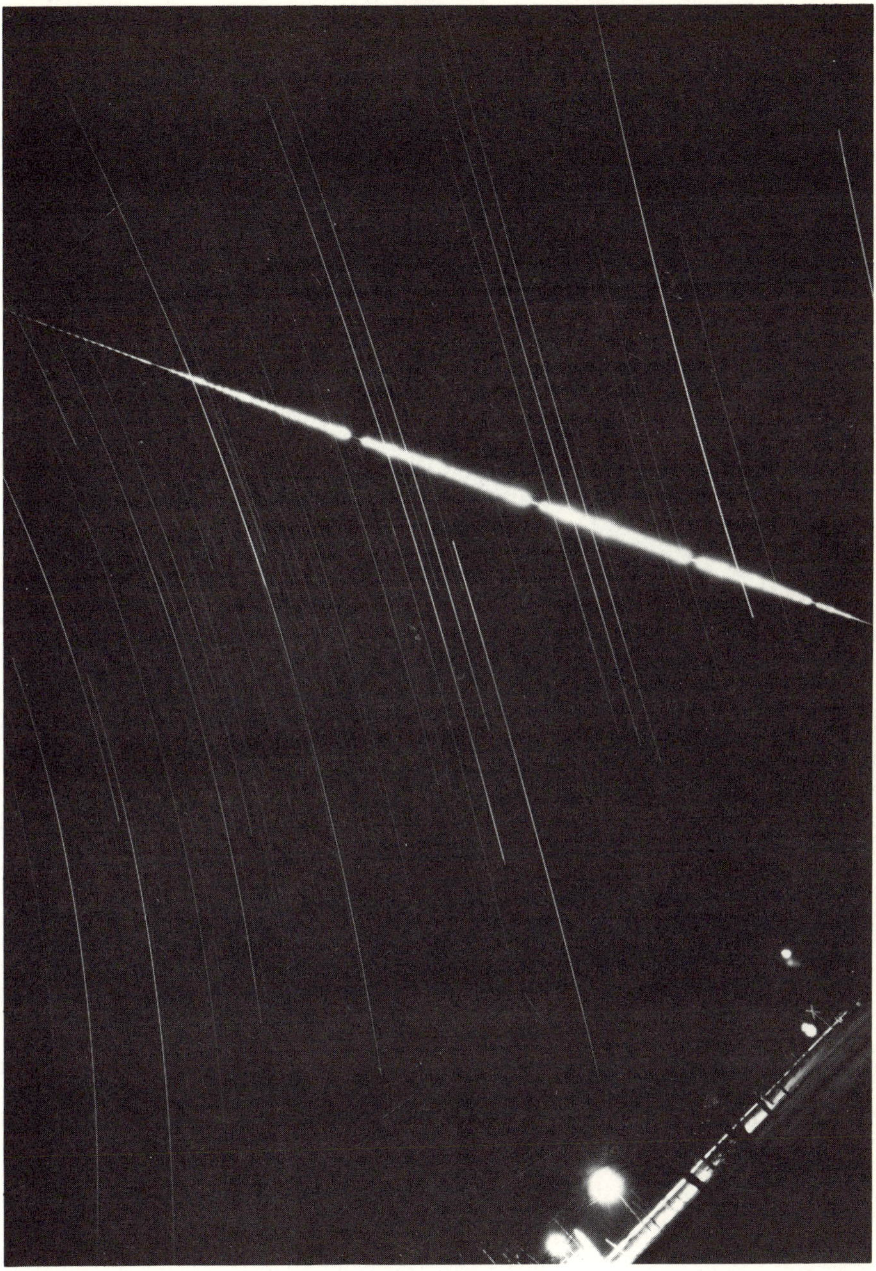

15. A railroad crossing sign clearly illuminated by street lamps is ignored by the Lost City Meteorite as it completes its orbit at an estimated two miles per second—a record low for a fireball in the earth's atmosphere. Descending over eastern Oklahoma on January 3, 1970, it belied its name when the first specimen was found within a week. Subsequent finds brought the total mass recovered to more than thirty-five pounds. It was the first meteorite whose impact point was calculated from the photographically determined trajectory. *Photo: Smithsonian Institution.*

16. The fossilized remains of this probable ancestor of the elephant, discovered in Missouri, are a testament to the great mammals which once roamed the North American continent, surviving the ice age only to die out with the rise of man. The mastodon shared the plains with the horse and the camel, among others. The horse was successfully reintroduced by the Spanish colonizers of the new world, but a nineteenth-century attempt to re-establish the camel as an inhabitant of the far west met with failure. Today, the mastodon's closest descendant—the elephant—exists tenuously in America by courtesy of the circus and the zoo. *Photo: St. Louis* Post Dispatch.

17. A woman of the Tasaday Forest people. *Photo: PANAMIN, Manila, Republic of the Philippines.*

18. A child of the Tasaday Forest people. *Photo: PANAMIN, Manila, Republic of the Philippines.*

19. Two oceans and a continent separate the Tasaday from another Stone Age people, the nomadic Akurijo of Surinam, South America, who were contacted at approximately the same time. Here an Akurijo woman uses the same kind of cutlery as her distant Tasaday contemporaries, a bamboo knife to prepare a spider monkey for the cooking pot. *Photo: Art and Evy Yohner, West Indies Mission, Paramaribo, Surinam, South America.*

houses and cars. A recording was made of the sonic boom by a girl from Bangor, County Down, who was taping bird calls at the time the bolide was seen and caught this fragment of the music of the spheres instead. Also reported from various areas—Dublin, Belfast, Antrim, Lisburn, Dungannon, and Kilrea—was a "swishing" or "rushing" sound as the fireball passed over.

The most constantly observed color was blue-green over Wales and "fiery white" in Northern Ireland with an intensity equal to, or brighter than, the full moon. Other colors seen were red and orange.

Fragmentation was clearly seen by a number of people. Near Aberystwyth the meteor broke into approximately six pieces which extinguished in a fraction of a second. Other reports of fragmentation came from Dublin and several other points on the trajectory. In one region, this action was described as a spray of bright red particles which seemed to be breaking up.

"It Was Warm but Not Hot . . . in One Nice Piece"

(The Haverö Meteorite Fall: Finland, 1971)

At 3:45 P.M. on August 2, 1971, a meteorite fell through a roof on the island of Haverö, Finland, fifteen miles southwest of the town of Turku (Abo), in the southwest of the country. The building is a storehouse about a hundred yards from the main house of a farm.

The meteorite made a 6-inch hole in a sturdy roof consisting of concrete tiles and 1.5-inch boards and then continued through the 1.5-inch wooden cover of a box. It was found ". . . in one nice piece" on the bottom of this box about five minutes after the fall. A nine-year-old boy playing in the yard showed the farmer's family where the stone had struck. It was said to be ". . . warm but not hot" at the moment of recovery. The barn was ". . . filled with smoke but was not on fire," which was taken to mean that a great deal of dust had been stirred up by the impact.

There were numerous reports of sonic phenomena, ". . . first like a thunderstorm, then like a jet plane, and then a big bang." These sounds were heard about one minute in advance of "the big bang."

The sun was shining brightly at the time, and there were no reports of light phenomena. The stone came through the roof almost vertically, as the alignment between the hole in the roof and that in the box showed.

The Haverö stone was scientifically classified as a Ureilite meteor on the basis of its texture. Ureilites are a rare type of meteorite which contain diamonds. Only four Ureilites are known, and there was preliminary evidence of diamonds in the Haverö stone.

A Sound as Loud as an Airplane

(The Police Meteorite Fall: Czechoslovakia, 1969)

On September 16, 1969, at 7:15 A.M., Greenwich Mean Time, a bright fireball signaled the fall of one of the year's most significant meteorites.

Witnessed by five persons, the object crashed into a farmhouse located in Suchy Dul, near Police nad Metuji, Czechoslovakia. The object smashed six roof tiles, destroyed a roof support beam, and broke into two pieces. One piece, weighing 2 ounces, fell into the loft, while the other, a 1.6 lb. specimen, rolled down the roof into the front yard. The owner of the farm reported that a sound as loud as an airplance accompanied the fall.

The height of the meteorite as it started its fall was between 16 and 17 miles; the height at the point of burnout was almost 12 miles.

The larger of the two pieces of stone chondrite measured approximately 5 by 2¾ by 2 inches. The estimated fallen mass was more than 1.8 lbs.; the mass found was 1.7 lbs. Approximately 4.5 ounces of the stone were distributed among scientific institutions in the United States, Germany, and Czechoslovakia for such diverse purposes as mass spectrometry, radiation analysis, and mineralogical investigation. The Smithsonian Astrophysical Observatory was included in the recipients.

The remaining 1.5 lb. piece was placed in the National Museum in Prague.

The Moon Is Not Immune

(The Lunar Meteorite Impact of May 13, 1972)

A huge meteorite crashed into the moon with tremendous force on May 13, 1972, and the Manned Spacecraft Center in Houston said it landed uncomfortably close to the nuclear-powered Apollo 14 science station.

The impact, equal to about 1,000 tons of TNT, was recorded by four seismometers placed on the lunar surface by various Apollo missions.

According to a Houston scientist, the meteorite made a crater as large as a football field on the moon, "a few tens of miles" from the science station on the moon's Fra Mauro formation. He said that the strike caused a rain of debris that lasted about a minute.

There was no indication that the science station was damaged by the flying debris, but dust may have collected on the instruments.

A White Cloud and a Flock of Black Birds

(The Kiffa Meteorite Fall: Mauritania, 1970)

In early 1971 the Center for Short-Lived Phenomena received a report from Mauritania of a meteorite fall which had taken place in October of the preceding year. It is perhaps one of the most informative accounts ever submitted on an event of this kind, and is presented in its entirety, without any changes or omissions from the original.

Description of Event

The following is the report of investigation conducted by the Management of Mines and Geology on Saturday, October 31, 1970:

On Friday, October 23, 1970 at about 2:55 P.M. local time (and G.M.T.), a high altitude explosion was heard in and around Kiffa.

A meteorite fall was recorded at 11°20′ W and 16°35′ N.

Thanks to the diligence of the Kiffa Department Prefect, Mr. Mohamed Baba Jal, aided by the gendarmerie and local authorities, an investigation could be conducted immediately and the specimen recovered and transmitted almost completely intact to the capital.

An investigation conducted by the Management of Mines and Geology on Saturday, November 7, was made possible by the reconnaissance organized by the Governor of the Third Region and by the kindness of parties interested in determining the conditions of arrival of the meteorite on the ground.

Two sets of testimony were obtained: the first at Kiffa, the second at the impact location of the recovered specimen.

I. *At Kiffa,* the regional livestock-breeding director is the main visual witness.

Noise: two loud explosions separated by more than a minute.

Seen: a white cloud from which emerged a "flock of black birds" (Chief of Gendarmerie). The cloud was seen vertically over Kiffa.

The primary direction of the cloud's elongation (which "was not a jet plane trail") and of the "flight of birds" provided a direction west-southwest to east-northeast.

II. *At the impact point.* An encampment slightly further away has been able to give us more precise information than the closest encampment, which appeared to have been very frightened.

The child of the nearby encampment who found the specimen about 10 A.M. the next day told us of noises which were heard, of a black cloud and black points, and also of a general N–S direction.

The woman who had first spoken of flashes and excessive brilliance has completely retracted her statements.

In the slightly more distant encampment, two large detonations were heard at a "small interval," the second being followed by "machine-gun chatter." The visible cloud was also on the vertical (at 8 km from Kiffa). It was white and one (or more) black point(s) emerged from it. This black point indeed seems to have been the recovered meteor itself.

The impact: A hole less than 20 cm deep in fine sand with no trace of burning. The "cram-cram" plants at the impact itself had been dispersed (buried) whereas the nearby plants had not even been scorched. The specimen must therefore have been relatively cool and brittle. It shattered itself into many tiny pieces.

The specimen: Leucocratic (light-colored) rock with very fine grains, filled with small points having a metallic luster. One is reminded of a feldspar rock having either bubbles or matrices of crystals with a cubic system.

Ashy texture (or inadequate binder), probably microcrystalline. The external part of the recovered specimen is composed of a black crust, probably due to fusion at a very small depth.

The rock density is relatively low for a meteorite, which therefore was from a sialic (rich in silica) group.

The total weight of the meteorite which reached earth must have been less than 2 kg (5 pounds), the sum of the recovered specimens being less than 1.5 kg.

Interpretation: A celestial body originating from an undetermined part of our universe, almost certainly from our solar system, was captured by celestial attraction resulting in entry of the specimen into the atmosphere accompanied by fierce heating, resulting in intense luminous phenomena.

The rocks are generally poor heat conductors and the phenomenon occurs rapidly. If the sample is sufficiently large, the nucleus has no time to undergo the heating effects which nevertheless volatilize its external layers. As a result, phenomena of contraction (fierce cooling) and expansion are super-imposed, leading in general to cleavages and cracks.

The noise (sonic boom) of its entry into our atmosphere could quite well have occurred at the end of, or after, the volatilization phenomena, but this is far from certain. On the other hand, at least partial splitting up did occur. The witnesses are, in fact, insistent on the existence of a white cloud (of dust and balls of steam) from which the "cold" non-incandescent specimens emerged.

Many small specimens must have reached the earth's surface, probably in the vicinity of the known impact (although this area could be immense); the fall of numerous small specimens was responsible for the "chattering" noise heard. It is quite possible that significant fragments fell out of range of any observer.

At least one meteorite explosion seems to have occurred south-southwest of Kiffa. A second explosion (with no cloud this time) or a contact with the dense atmosphere would have caused a great burning near the impact point.

From the moment of the visible explosion until the impact of the recovered meteorite, the mean speed of the specimen would have been slightly less than the speed of sound. However, the speed limit of such a specimen in the earth's dense atmosphere could hardly be greater than 400 kilometers per hour and the "shock" in the sand proves that this speed could not have been exceeded by much.

The crackling noises heard were certainly attributable to the practically simultaneous descent of the large specimen and the multitude of small fragments ejected in this direction by the main explosion.

Sifting of the nearby sand would probably yield numerous small specimens.

—M. P. Pellas

Dr. Pellas led an investigation sponsored by the Centre National de la Recherche Scientifique (C.N.R.S.) of France. He reported additionally that the meteorite is probably an H_4 chondrite (a form of stone clustered in granules). The texture is highly chondritic, and the chondrules are very easy to remove from the sample. It is an exceedingly friable stone.

A broken fragment of 1.3 lbs. was measured for cosmogenic radionucliotes. From this fragment, 5.7 ounces were taken for the meteorite collection of the Musée National d'Histoire Naturelle in Paris, with whose Laboratoire de Minéralogie Dr. Pellas is affiliated. The remaining 15 ounces were returned to the Mauritanian government.

Part IV

URGENT ANTHROPOLOGICAL EVENTS

A Stone Tool Culture

(Discovery of the Tasaday Forest People: Mindanao, Philippines, 1971)

In the summer of 1970 the private Philippine agency concerned with national minorities (PANAMIN) began hearing rumors of an unknown people in the heavily forested mountain region of Mindanao's southern Cotabato Province. Hunters living in the vicinity of the isolated Manubo Plateau (or Manubo Blit) were said to have glimpsed strangers in the forest and to have come upon camp sites where animals had been butchered and the pith of the wild palm prepared.

PANAMIN tried to confirm these rumors but met with only hearsay until their inquiries reached a Manubo hunter named Dafal. He had the reputation of "walking through the forest like the wind" and insisted that he had not only contacted the new people but had also traded with them and stayed overnight in their camp. According to Dafal, they called themselves the "Tasaday," after a mountain near their home. Further, they had absolutely no agriculture, no cloth, and no metal tools when he first encountered them. Their staple foods were the pith of wild palm and wild tubers.

The reported Tasaday habitat was probably the most isolated mountainous area within the Republic of the Philippines. It was all above 3,000 feet elevation, deep in an unbroken expanse of rain forest covering hundreds of square miles. The rugged terrain, bisected by small streams which cut dark, narrow lanes through the towering trees, inspires awe when seen from the air. It was difficult to believe that this vast and undulating sea of tropical growth could harbor man.

The description of what seemed to be a Stone Age culture in a trackless wilderness so intrigued PANAMIN that plans were made for a full-dress meeting with the Tasaday. Dafal volunteered to contact the group he knew and, if possible, to assemble them in a clearing on the Manubo Plateau at the edge of their forest home. This site had the additional advantage of being suitable for a helicopter landing.

Dafal went off into the interior and returned a week later with word that the Tasaday were agreeable. He was then airlifted back to arrange the actual contact for two days later, June 7, 1971.

On that date a helicopter landed at the rendezvous, carrying a PANAMIN research team headed by Presidential Secretary for National Minorities Manuel Elizalde, Jr. Dafal emerged from the forest followed by four frightened men, their fear heightened by the whirling blades and engine noise of the helicopter.

During the afternoon and the next day twenty additional Tasaday—women and children included—visited the temporary camp in the clearing. This same group returned for the second visit of the research team on June 16. They were the only Tasaday contacted on these first occasions, but they mentioned that there were larger numbers in the forest, and the helicopter pilot observed smoke from a camp site deep in the jungle.

A fascinating picture developed from these two initial contacts, and Elizalde's report reflects great pride in coming face to face with a living testimonial to Filipino prehistory. He couples this with a firm commitment that any change which may occur in the culture of the Tasaday through their new-found connection with the outside world will be with dignity and by their own choice.

Whatever impact the encounter may have had on the Tasaday, the impression they left in turn on the research team was equally overwhelming in its implications.

They were found totally lacking any knowledge of rice, taro, the sweet potato, corn, cassava, or any other cultivated plant, and had no names in their language for these cultigens. They had never tasted salt or, even more significantly, smoked tobacco. The tobacco plant, like corn and the sweet potato, is of American Indian origin and had literally swept the world in the seventeenth century. It had reached the most isolated peoples of Africa and New Guinea, but it had apparently not penetrated to the Tasaday, placing them among the few (if not the only) people in either hemisphere today who neither know nor use tobacco.

The Tasaday have a basic stone technology. They depend primarily on stone tools as scrapers, pounders, and chippers. The key to their ability to survive without metals is bamboo, which they fashion into containers for cooking and carrying water, as well as traps, blades, awls, saws, and other items of material culture to exploit their environment. This is the only positive record of a Filipino people using stone tools to make implements of bamboo for cutting, piercing, and sawing.

The Tasaday are true hunters and food gatherers who never made the transition to the agricultural society that followed the Stone Age. Instead of a settled existence marked by the cultivation of crops and the keeping of domestic animals, they still roam the forests from encampment to encampment in search of game and edible wild plants.

PANAMIN estimated that the separation of the Tasaday from their native parental stock, the Manubo, took place at a relatively early time. They still retain some of the characteristics of the basic Manubo culture and language. The similarities include elements of mythology, religious beliefs, and even the custom of tattooing. But the differences are considerable. They lead the PANAMIN researchers to believe that the Tasaday separated long before Spanish days in the Philippines (four to five hundred years ago), perhaps as long as fifteen hundred to two thousand years ago.

Although their language is distinctly Filipino, a member of the great Malayo-Polynesian family of languages, no one could be found among the neighboring peoples who could understand the Tasaday tongue well enough to elicit detailed information during the first contacts. A Manubo Blit woman who spoke Tboli was used as an interpreter of Tasaday, but at least half the words were unknown to her. What she could glean was translated by a Tboli who also spoke excellent English and Tagalog.

Among the words which are completely lacking in the Tasaday vocabulary are those for "sea" and "boat," their world being circumscribed by the rain forest. They neither use, nor have they heard, the terms Filipino, Moro, Tboli, Ubi, or like designations for other peoples. Their social and geographical isolation is almost complete, although they have sometimes heard and seen hunters in the forest, and speak of the fields and houses of people who live at lower elevations—toward where the sun sets. The knowledge that there are people around them is terrifying, the Tasaday freely admit, and they did not ask either Dafal or the PANAMIN researchers a single question about where they had come from at any point. They were even reluctant to talk about other local groups of Tasaday and identified only one other, larger than their own, which they called the Sanduka. According to PANAMIN, the Sanduka and the known Tasaday together would probably number about 60 men, women, and children; the total Tasaday population, it is felt, does not exceed 100, since the number is restricted by their technology and the capacity to exploit the resources of the forest with their primitive tools.

The great fear of the outside world which the Tasaday display may have a practical basis. When questioned about what they fear most, they replied that it was not evil spirits or even the inexplicable aircraft that fly over their forest domain; it is *fugu,* or epidemic sickness.

The life span of the Tasaday is obviously short. No elderly men or women were found in the group contacted. There are memories of a smallpox plague—*fugu*—which swept the area. Sickness resulting from external contacts would explain the absence of older individuals, as well as the pervasive fear of outsiders and the world around them. It is likely that the Tasaday have not acquired resistance to any kind of infectious disease and that their extreme isolation, while it is the only real protection they have, at the same time prevents them from developing any degree of immunity.

They are nonetheless a fertile people; out of the 24 individuals in the local group, 13 are below the estimated age of ten years, including still unnamed infants. One child, a retarded male albino suffering from a terrible skin rash, was hazarded no chance of survival by the PANAMIN team. In their mobile state, without permanent shelters to ward off the cold and rain of the high-altitude forests, the Tasaday can offer only a precarious existence to their weak and aged. Those who die of old age or accident are simply buried in the ground close to the spot. Sick people, on the other hand, may be abandoned to die alone without help or medication, according to one member of the group. If true, this is apparently out of fear of epidemic sickness, of which there are vivid recollections.

The social group in which the Tasaday hunt and gather food is called a *nasagbung.* It is composed of a number of distinct elementary families, or *setalawmin,* made up of father, mother, and unmarried children. Orphans, widows and widowers without children attach themselves to one of these nuclear families.

For sleeping and protection, each family builds itself a shelter out of palm leaves, crudely plaited for flooring and propped up by branches to form walls. For more frequently used sites, rock shelters are sometimes constructed.

The family is the basic economic and political unit, although it joins with other families, when advantageous, to hunt and gather food. PANAMIN is careful to describe the social grouping of a number of families, or *nasagbung,* as a "local group," rather than a "band," since there is no formal leadership. Decision making, which is primarily concerned with food-getting activities, is based on the age and greater experience of individuals and upon the consensus of the adult men and women who generally share opinions.

In hunting, the Tasaday use the *balatic,* a method known probably to all Filipino peoples. It is a bow device which releases an arrow when the game strikes a trip cord and is unquestionably the most effective trap for bagging wild pigs and deer.

Despite their use of the *balatic,* the Tasaday had no knowledge of the bow and arrow until Dafal gave them five Ubo bows and thirty iron-tipped arrows on his second visit in 1967.

Among their tools, stone used for scraping is made from cryptocrystalline quartz, while river cobbles serve for smashing and pounding. The use of the varied bamboo implements shaped with stone scrapers can be seen in the preparation of a favorite dish, monkey meat. The hair is first singed from the animal, and a bamboo knife is utilized to cut between the joints of the arms and legs so that the limbs can be broken off. The body is then quartered with a sharp bamboo blade, and the meat is either roasted over an open fire or boiled with water in a bamboo cooking tube.

Supplementing bamboo as carriers, rattan and black vines (*Lygodium circinnatum*) are woven into baskets. The bark of trees is used for cloth, as it is elsewhere in the Philippines. This is done simply by beating the trunk of a tree, mainly the wild fig, with a heavy stick. The outer bark is then torn away from the trunk and peeled away from the bast. Next, the bast is dampened and pounded at the bank of a stream with any river cobble that happens to be handy. After this it is pliable and can be trimmed with a bamboo knife. The cloth is worn on T-strings woven from the broad leaves of the ground orchid. One of the Tasaday adults said that, unless this rudimentary covering was kept on, their organs would be vulnerable to witches.

For ornamentation, both men and women wear a number of brass rings attached from the helix to the lobe of the ear. Most of the brass wire now used by them for these earrings was a gift from Dafal, but they also pointed out that they had inherited some old earrings from their ancestors. Whenever the wire for these prized ornaments was obtained, it must have been in the remote past, since the present-day Tasaday group insisted that Dafal had previously been their only link with the outside world.

The sole musical instrument known to the Tasaday is a jew's-harp, or *kubing,* which they keep in a bamboo container and carry from place to place.

Getting married is not easy for the Tasaday, owing to the small size of the group and the imbalance in the number of boys and girls of marriageable age. In the local group contacted, there were 9 male to 4 female children. Even so, the Tasaday are monogamous; they do not practice either polygamy or polyandry, although this is common among other known hunting and food-gathering peoples in the Philippines. There would certainly be no moral reason for prohibiting polygamy, since the ancestral creator of the Tasaday, *Bingbang,* whom they share as a deity with other Mindanaon people, is said to have two wives.

One of the young Tasaday men, a personable twenty-year-old orphan named Balayem, expressed great fears about his future role as a husband and father because of the lack of eligible girls. When the PANAMIN team suggested jokingly that they might help him find a wife, he took their offer most seriously and reminded them of it on their second visit.

There is some evidence, which remains to be checked, that the Tasaday indulge in "bride capture," the forcible taking of women from other local groups. One woman said that she was ". . . alone in the forest of Sanduka when her husband's father found her." The inquiry was not pursued, since the father-in-law in question was clearly disturbed at the remark, and it is believed that the reluctance of the Tasaday to speak of other local groups stems partly from the practice of abducting brides.

Beginning in August 1971, the PANAMIN research staff undertook a prolonged study among the Tasaday. During this time they were asked not to use any of the material culture obtained from Dafal or from the team. An inventory was made of all items which came into their possession from these sources, since even the presence of one broken metal bolo, given to them in 1968 by Dafal, represents a tremendous technological improvement.

In studying the ability of the Tasaday to survive without agriculture and without metal in the midst of a tropical rain forest, PANAMIN can write one of the most fascinating chapters in the history of mankind. Among the principal questions are: What food resources does the forest have, such as plants, animals, fish, crustaceans, and insects, which the Tasaday gather, trap, and hunt? What are the tools they use in their gathering and hunting activities, and how are they made? What quantity of food is necessary for survival in one area before resources are exhausted? What is the relationship between the wild food resources of a given area of the forest and the number of people this area can support? For how long? How do the Tasaday compare physically with lowland agriculturists? With slum dwellers of Manila?

Aided by a botanist and a zoologist from the Philippine National Museum, the PANAMIN field team will live with the Tasaday during different periods of the year, contrasting one season with another, to reconstruct the metal-less culture and society of these unique people. They provide an unparalleled opportunity to gain an understanding of man before the appearance of agriculture and the domestication of animals; and before the appearance among modern men of a highly complex technology based on metals. It is this latter-day technology which is devastating the environment upon which man everywhere once depended, and upon which the Tasaday still depend.

There is an immediate threat to the survival of the Tasaday in the plans being made by loggers to drive roads into their great forest sanctuary. In addition, neighboring slash-and-burn agriculturists are destroying many thousands of square miles of the climax tropical rain forest in which the Tasaday roam.

PANAMIN, which is engaged in a continuing campaign for balanced conservation on behalf of national minorities, regards the Tasaday as a living symbol of the struggle, for when the forests disappear, so will they. The agency is attempting to organize a combined effort by the public and private sectors of the Republic of the Philippines to institute a sound program of land preservation, utilization, and allocation, together with viable programs of social and economic development among all the varied peoples of the nation.

The Tasaday way of life will not long survive contact with the outside world, and PANAMIN is determined that these forest people will have the maximum control possible over the kind of change they undergo and the pace at which it takes place.

On Dafal's first five trips, he made gifts to the Tasaday of bows and arrows, spears, metal blades, some pieces of cloth, and a few miscellaneous iron objects.

On their two visits, the PANAMIN team gave similar items, mainly of Tboli manufacture, plus three mirrors and a flashlight. They also brought foodstuffs, including rice, sweet potatoes, salt, sardines, biscuits, boiled eggs, limes, candy, sugar, and dried fish. They did not forget tobacco.

The Tasaday at first tried to eat the rice uncooked and were distrustful of both salt and sugar. "Perhaps it is poisonous," one man shyly ventured when given sugar.

On the second visit, however, the children were clamoring for more of the candy which had been brought them nine days earlier. And their elders showed themselves, at least in one respect, to be somewhat in advance of the thinking of a good part of the civilized world. They returned the tobacco.

The Tasaday were left to continue their existence undisturbed until the spring of 1972, when Manuel Elizalde led a party of anthropologists and journalists into the Manubo Blit for the first contact since the previous summer's. They soon experienced one of the principal problems of the people they had come to visit—the necessity of gathering food. Having run out of provisions and lacking both the resourcefulness of the Tasaday and their taste for such protein staples as large tadpoles, they were airlifted out. Upon emerging, one member of the expedition said, "It is almost unbelievable—a shock like going back into time thousands of years." The group had discovered, among other things not known as a result of the initial encounter, that the Tasaday did not live in crude huts in the jungle—these were apparently only temporary shelter on hunting trips—but in limestone caves 600 feet up a mountainside. It was, as another member of the group said, a major scientific discovery. For the first time, anthropologists would get a chance to study primitive cave-dwelling people in the flesh.

Shortly thereafter, President Ferdinand E. Marcos, of the Philippines, proclaimed about 50,000 acres of mountain forest in Cotabato Province the exclusive reservation for the 24-member Tasaday tribe and the slightly more advanced and more numerous Manubo Blit tribe. The area cannot be leased to loggers, whose encroachments were threatening the area even before the disclosure that it was the home ground of the Tasaday.

The President also reappointed Manuel Elizalde, Jr., as his assistant for cultural minorities, perpetuating in office the man whose first reaction to the Tasaday was that their way of life would, at all costs, be preserved.

Mastodon from Missouri

(A Rare Bone Find, 10,000 to 30,000 Years Old: Missouri, U.S.A., 1969)

In the last half of the twentieth century, in southwest St. Louis County, Missouri, an area designated for the construction of a supermodern office complex, the driver of a bulldozer uncovered the bone of a mastodon, a creature that roamed the earth 10,000 to 30,000 years ago. The bones were 23 feet below the grade and were excavated by a research team of archaeologists while construction material was removed from the area.

Mastodon is the name given to those extinct mammals and probable ancestors of the elephants in which "the teeth are of browsing type, composed of low crests and nipple-like cones rather than the grazing type composed of high, flattened plates" that mark the mammoths and elephants themselves.

The Oligocene Epoch, which began thirty-six million years ago and lasted about eleven million years, saw the increasing emergence of North America. Mammals were dominant then with cat and dog families supreme. The three-toed horse as well as the mastodon was present. Primates walked erect and the primitive anthropoid ape made its appearance.

The earliest known forms of the mastodon associated with the Oligocene Epoch in Africa were approximately 4.5 feet in height. Their descendants, the size of large elephants, spread over Eurasia and North America, persisting into the Pleistocene Epoch. The Pleistocene Epoch, or Ice Age, began about a million years ago. It was then that glaciers covered parts of the Americas, Europe, and Asia four or five times and that four species of elephants, including the mastodon and mammoth, the camel, and the sabertooth tiger were present. Mastodons occurred on all the continents except Australia. In a more particular sense, the name is often used for the commonest American mastodon of the species, frequently called *Mastodon americanus.*

Discovery in the eighteenth century of the remains of the common

American mastodon had a great deal to do with the development of the science of paleontology. These remains have been found in deposits of the Pleistocene Age throughout the United States and Canada though they are considered to be rare in the Middle West and southern United States.

In life, the animal closely resembled an elephant and had similar tusks and a trunk, but was lower and longer in relative proportions. The grinding teeth are readily distinguishable by their pattern of low transverse ridges. It is evident from this tooth structure and from the frequent occurrence of its remains in swamp, bog, and stream deposits that this species lived on leaves and preferred the environment of the wet, lush Ice Age forests.

The immediate ancestry of the species can be traced back to the Middle Miocene in North America and the Early Miocene of the Old World from where it migrated to America.

The remains found in Missouri were reported, by the curator of the St. Louis Museum of Science and Natural History there, to include "several vertebrae, some ribs, long bones, parts of tusks, scattered bone fragments and also parts of teeth."

Because of the teeth, the archaeologists were able to determine that the bones were those of a mastodon. In addition to the bones, traces of charcoal and two tiny snail shells were found at the same level.

There is no recorded instance of a mastodon who had his foreleg made into an umbrella stand, tusks carved into back-scratchers, or who marched in a circus procession with other mastodons. Those honors were reserved for his descendants, the elephants, who had the misfortune to be contemporaries of Neolithic man. Neolithic man, in turn, has descendants who speculate on the reasons for the extinction of the great land mammals who were once so numerous.

A Camel Covers Its Tracks

(Santa Fe Prehistoric Camel Tracks: New Mexico, U.S.A., 1969)

Somewhere between seven thousand and a hundred and fifty thousand years ago, a solitary prehistoric camel, one of the contemporaries of the American mastodon, ambled across a patch of mud thirteen miles southwest of what eventually became Santa Fe, New Mexico. Leaving imprints of a set of outsize hoofs, it then meandered off on the road to continental extinction. A nearby cluster of small but active volcanoes obligingly emitted a mass of hot cinders while the tracks were still fresh, effectively burying, baking, and preserving them in a single operation.

A geologic era passed; the volcanoes ceased huffing and puffing; the mud turned to stone. The age of the bulldozer arrived.

As 1969 dawned, a power shovel operator, scooping out cinders for a manufacturer of building blocks, noticed a line of regular depressions in the layer of soft stone under the cinder bank. He stopped and notified his employer, who called in experts from the University of New Mexico. They immediately identified the find for what it was and consulted with the Federal Bureau of Land Management. The Bureau, which has jurisdiction over the land, withdrew it from use, leased the cinder-block manufacturer another pit, and built a 100-foot shed over the discovery.

Clearly the outstanding example of locking a stable after the occupant has vanished, the shed has a solid roof and mesh sides through which the public can view, but not touch, the procession of petrified camel tracks. And, appropriately, it also displays a large sign that describes what is inside and gives due credit to the machine-age man who uncovered the Pleistocene pad marks.

A Bloodstained Stone—Ritual Child Sacrifice?

(Discovery of the Guajaki Indian Tribe: Paraguay, 1970)

A group of Stone Age nomadic Guajaki Indians was discovered during the course of a forest inventory by the Food and Agriculture Organization of the United Nations together with the Paraguayan Ministry of Agriculture. Of additional significance is the fact that the culture and perhaps the lives of the people are endangered by the expansion of logging and agriculture in the region.

This group of Indians was discovered about forty kilometers west of Paranambu on the Río Paraná in Paraguay on March 30, 1970.

Members of the inventory field crew brought from the forest a number of artifacts found in huts that had been abandoned only shortly before the crew's arrival. Certain members of this group then reported to an internationally known ethnologist whose special field is the Guajaki tribe. In his opinion, the tribe had almost certainly lived virtually without contact with Europeans. Since there are few Guajaki remaining in their native state, and since there is the possibility of the group being driven forcibly from its area as a result of rapid economic expansion, he believed a scientific study to be a matter of utmost urgency. He also felt it absolutely essential to avoid attracting attention or publicity until the study is concluded. Consequently, at present, only brief notes on the discovery of the tribe are available.

This material included the following information. There were observed two pits for trapping tapirs. In the Guajaki encampment there were eight fires but no huts except "Tapy yva"—an enclosure used for the ceremony of perforating the lips of young boys. There was also one "Tapy Jyvapa," a shelter for a young girl experiencing her first menses. Shavings of what was evidently "kymata" (a kind of liana), used for the ritual bath, were found too.

The notes on discovery also recorded: "Short ravé/chavé (scarification) used evidently to test boys whose lips were perforated—bloodstained."

The concluding notation reads as follows: "Two stones—one bloodstained. Ritual sacrifice of child? 35 to 40 kilometers west of Puerto Ñacunday, Río Paraná, Paraguay."

Nomads with Stone Axes

(Discovery of the Akurijo Indian Tribe: Surinam, 1968–71)

The earlier discovery of another Stone Age tribe, the Akurijo, of South Surinam, inevitably invites comparison with that of the Tasaday people, half a world away.

In the case of the Akurijo, a prolonged period during which they could be studied without the incursion of outside technologies or culture was a luxury they could not afford. They were in dire need of rescue from an environment that could no longer support them, and the missionaries who made first contact, with the concurrence and aid of the Surinam government, saw that the only salvation for the Akurijo was to join them with neighboring Indian tribes who had already made the transition to a settled agriculturally based life.

If this would change the Akurijo tribally, perhaps beyond recognition, it would also save them as individuals with some chance of preserving their identity as a people after their futures were secured. It is in this light that the account which follows should be viewed.

Missionaries carrying the word of God to unknown peoples deep in uncharted jungles are not a dead chapter out of the romantic past. The story is still being written today in our hemisphere, sometimes with dramatic encounters reminiscent of the age of exploration.

For three years missionaries of the West Indies Mission had been searching for an isolated nomadic stone-ax tribe in Surinam, South America. In June 1968, while on a mapping expedition up the rivers of the mountainous area separated from the rest of Surinam by its topography, they made their first contact with one group of this tribe. Since traces of these primitives had been seen at various points on the rivers during the previous three years, the missionaries hoped, indeed expected, to find them eventually. However, it was with some surprise that they discovered they were being watched during the entire three-month period the mapping project was in the area. Although their tracks had been observed and even a bit of beeswax bearing an impression of human teeth was found, the group managed always to stay out of sight and it was by accident that the initial contact was made.

As the main body of boats of the mapping group appeared to have gone by, a group of watching males neglected to conceal themselves; thus, when one canoe, following the rest and moving very slowly along the riverbank, came upon them, they didn't have a chance to run.

It was at this time that these primitive Indians were thought to be part of the Wama tribe originally located in 1938.

On September 9, 1968, the missionaries and fourteen Indian helpers left the Lawa Mission post to try to relocate and contact this elusive people for the second time. They made their base camp at the mouth of the Waremapankreek and the next day proceeded to the point where the Indians were first seen in June. From there they trekked into a camp previously used by them but found it deserted and showing no signs of recent use. Seven of the Indian guides were sent out on the trail to the camp in which the missionaries had spent the night with them in June, in order to determine how long ago the people had evacuated the camp and in which direction they had gone. Apparently they had pulled out immediately after the June contact as they had indicated, and it soon became evident that there were no members of this tribe in the Waremapankreek area. Except for one of their very crude huts several weeks old, in which one member of their group had apparently spent a rainy night, neither did the rest of the party find any recent signs of the Indians as they poled the canoe farther up the Waremapankreek.

It was presumed the group had gone west, parallel with the Litani River. So the missionary party proceeded west, too, hoping to intercept them. The river was rapidly reduced in size to a creek and it was necessary to chop their way through the many trees fallen across the river. While the missionaries proceeded upriver chopping trees in order to get the canoes through, several of the Trio Indian guides cut a trail south in search of their trail. They found several hunting trails but not *the* trail west.

Suddenly they came upon a bridge obviously made by this tribe. It was a large tree fallen across the river with a vine handrail stretched beside it. Along the trails on either side of the river leading to and from the bridge were twigs and saplings broken over by hand, obviously several months before. There were stone ax marks on the trees but no evidence of the machetes and axes given to the Indians on the first contact. It was deduced from this evidence that this was the trail by which the Indians had come east *into* the contact area. So it was decided to continue upriver in search of a similar crossing to the west.

The keen eyes of the Indian guides in the party noticed several fishing spots on the banks of the river. These were important and encouraging as they were more recent signs. A fishing spot consisted of a strategically placed palm frond to shield the fisherman from the view of the fish, and a forked stick used to suspend entrails tied to a vine at the water's surface. As the fish surfaced for the bait, it was shot from behind the blind with bow and arrow.

Late in the afternoon they found their way blocked by an enormous tree across the river and simultaneously ran out of gas—having had to go farther than anticipated and hampered by a faulty outboard motor. Here, too, was a vine handrail strung along the length of the tree and intended for the Indian women carrying heavy back packs and for the children to use to steady themselves as they crossed on the great log. They immediately checked the trail to the southeast as it was the direction from which they presumed the group would be coming. It had been cut with *machetes* and *knives!*

Following the trail for exactly one mile, they came upon a deserted camp exactly like the ones seen in the Waremapankreek area, and all the signs were recent! A footprint, a piece of freshly cut palm bark from which a drinking container had been made, and some crushed poisonous vine in the swamp were certain indications that someone had been there poisoning the fish that very day, or the day before.

With darkness approaching, the missionary group returned quickly to the canoe in order to make camp on the north side of the river. They hastily built huts roofed over with palm leaves and settled in, planning to check out the trail on that side of the river in the morning.

Next morning, Sunday, September 15, some of the mission group returned to the deserted camp to look around again, while several of the Trios proceeded along the trail to the north. At 9:40 A.M. the Trios walked into another camp and came face to face with the group for which they had been searching. They were sixty miles to the west of the place of the first contact in June.

When the missionaries heard the news on their return to camp, they stopped only long enough to pick up a few trade goods and hurried on to join the Trios and make a second contact with the Indians who, they now learned, called themselves the Akuri.

Their camp was identical to all the others they had seen, eight huts built in a Wapu palm swamp. There were 15 Akuri Indians there—3 men including the old chief, and 12 women and children. One of the men and a young boy were completely new to the mission group while all the rest they had seen on the first contact three months earlier and many jungle miles to the east. This made a total of 30 individuals encountered in the two initial contacts—9 men, 8 women, and 13 children. And they were, indeed, heading west.

The area of 10,000 square kilometers where these non-agricultural people live is differentiated from the surrounding topography by the precipitousness of the mountains, so that agriculture is much more difficult. It had been the plan of the Surinam government, therefore, to send Bush Negroes into the center of the area to establish a manioc (cassava)-producing farm with the expectation that the Akurijo groups would then come into this settlement for trading purposes—to obtain trade goods and cassava.

Obviously the Akurijos did know of the existence of outsiders, for they had a few items of material goods probably picked up from earlier expeditions and/or from Trio and Wayana groups who could swiftly travel through the ordinarily impenetrable areas to Brazil.

The Akurijos have a tradition that the Wayana Indians killed many Akurijo ancestors a long time ago, and the Wayanas have a similar tradition. Hence the Akurijo tend to accept the Trio Indians somewhat more easily than the Wayanas. In Surinam, generally, there is a belief that wild and evil Indians inhabit the mountainous area. Consequently, all these beliefs have tended to keep the Indians apart from fear of each other, and though the Wayanas and Trios travel through this area to Brazil, they apparently stay on the beaten trails as those individuals known to the missionaries claim never to have seen any Akurijo mountain people before June 1968.

Akuri means "agouti," a South American rodent. "Jo" means "people," so Akurijo means "the agouti people," so called because of the importance of the little animal in their lives. The Akurijo eat agouti, including the marrow of the bones, which they crush for this purpose. They also use the jawbone of this rodent as an instrument for sharpening things. This, by the way, is not an unusual name—many South American tribes are named after animals, fish, and so on.

It became obvious to the mission group that the Akuri trails were very old and that the people traveled them with some sort of regularity, revisiting the same areas every couple of years, for all the camps showed evidence of having been rebuilt on several previous occasions. Although they are forced to move continually in search of food, they do so unhurriedly. Their camps are spaced from three to five kilometers apart and it was learned that the Indians may spend from a few days to three weeks in them, sometimes passing over two or three camps before stopping, with the group frequently splitting up to forage for food.

The mission group found the Akuri to be continually dirty. They never saw them bathe and it appeared that the dirt stayed on them for days. It seemed to be a result of their camping habits. They always set up a camp in a relatively dry flat area in the middle of a swamp, so that getting to and leaving the camp required wading through swamp waters. Apparently this is done not for defense but because it is the only relatively clear and flat area in this kind of country. In fact, there seemed to be no fighting at all among the various bands of this tribe and, in their unconcern for defense, they make no effort to conceal their trails.

The Akurijo skin color was described by the missionaries as pallid or sallow and attributed this to both diet and the fact that they seem so little exposed to the sun except perhaps only in the middle of the day. This because they make no clearings in the jungle and because the streams are largely forest-covered, too. Except for the color of their skin, they are very similar in many of their physical characteristics to the Trios and Wayanas. Although, in contrast to the Wayanas, who remove all facial hair including eyebrows and eyelashes, the Akurijos do not pluck out *any* of their facial hair.

The missionaries had discussed the question of giving trade goods as gifts or trading on the first contact with the Akuri and decided it was essential to do so at this time to express good will. Since the Wayana and Trio Indians went from place to place trading their own things, even their wives, as part of the exchange situation with the other groups on these trips, it was felt that it would be impossible to contact such a tribe and maintain any kind of a decent relationship with them if gifts were not made and barter encouraged. Indeed, it was felt that trade was expected in any sort of a novel contact.

The meeting with the Akurijos on the first day of this second contact was a good one. They remembered the mission group from the first contact in June and said they had heard them coming up the creek the day before. It was assumed, though they did not admit it, that they had also watched the expedition. Talk, trading, and a religious service followed, during which they managed to steal one of the Trios' machetes. A key factor in whether they wanted objects from the outsiders (missionaries, Wayanas, Trios), it seemed, was whether they could recognize the usefulness of the object to themselves. For example, at first they did not want pots, but as soon as they recognized that an aluminum pot was very much like one of their clay pots, only better, they were keen for aluminum pots. Hence they did not want pots during the first contact but they did on the second—though they threw away the lids. Also they have never wanted cameras or the red cloth of the Wayanas, though they wanted beads and axes right away.

Toward late afternoon on this first day of contact in September 1968, the mission group went back to their own camp, having promised to return to the Akurijos the next morning. During the night they shot two alligators which they gave the Akuri as a gift the next day. The women readily accepted the alligators and immediately began to prepare them in typical Indian fashion. A gift of iguana eggs, however, was thrown away, although several years ago traces were found of one band of these non-agricultural people living in this same mountainous region who apparently did eat iguana eggs. This is perhaps explained by tabu differences between the bands.

It also appeared that the Akurijos were scared by items of equipment which were not evident to them when the expedition first entered the camp. For example, they were not frightened by the tape recorder, which had been visible from the first, but were scared of the microphone, which they had not seen, and later scared by the movement of the tape recorder when it was turned on.

Before the mission group arrived at the Akuri camp on the second day of the September contact, many of them had gone off in search of honey and nuts. Since food is the essential drive of these hunters and gatherers, it was difficult to hold their attention for more than a few moments at a time. Because of their preoccupation with satisfying this need, they were constantly jumping up to crack nuts or to scan the trees for the possible activity of honeybees.

Bearing on the question of whether the Akurijos had lost agriculture at some earlier time or whether they were the remnants of an aboriginal Carib group existing in that very region since pre-agricultural Carib days was the statement by one Akuri woman who said that her mother had told her about cassava bread, though the Akurijos in general had never seen or heard of it. It was also interesting to learn that the same word, *ulu,* used by the Wayanas and Trios for cassava (tuber) and cassava bread, is used by the Akuri for the nuts upon which they depend so greatly. Nuts provide the staple for their diet as cassava does for the Wayanas and Trios. Apparently the Akurijos move from place to place searching continually for more honey. This appears to be an item of scarcity whereas their other foods, such as fruits, nuts, and meats, may be found in abundance. It also appears to be the one sweet item in their diet except for certain jungle fruits during the rainy season. The Akuri show no signs of malnutrition; they appear strong and apparently have enough to eat, lacking only certain minerals and vitamins.

NOTE: One of the mission group suggested, upon inquiry, that the Akurijos could have severed themselves some generations ago from the mountainous region, losing agriculture because of destruction in warfare.

On that second day of contact in September 1968 the mission group "talked with the few Akurijos remaining at the camp, sang for them in the Trio language, wrote down many of their words, and told them of God." While they were thus engaged, a few Akuri slipped away to the mission camp at the river's edge and stole several pots, knives, and some of the things they had traded away to the Wayanas the day before. The guards left at the camp had been unable to control the situation. The people apparently were not content with what they had been given or what they had gained in trade but were intent on possessing everything they saw. Since many of the stolen things were necessities for survival, a real effort was made to retrieve them. But these efforts were met only with the response, "We are good. We did not steal anything." Yet a search in the jungle adjacent to the camp produced one pot, a knife, and some bananas. While they were looking around, one Akuri partially drew an arrow on one of the Wayanas who had his back turned. Perhaps he was merely toying with his weapon, or perhaps he intended to shoot the Wayana, who was about to uncover one of the stolen articles. As one of the missionaries deliberately and noisily hurried through some dried leaves in the Akuri's direction, he withdrew his arrow.

Although these people took advantage of every opportunity to steal from the mission group and from the Indians in their party, they seemed less uneasy than at the first contact, for they carried their weapons only when they came to the missionaries' camp on the river.

On the first contact in June, they had refused to eat any of the missionaries' foods. In September, however, they greedily ate bananas, sugar cane, and cassava bread, although it was immediately obvious that these foods were completely unfamiliar to them—in fact, they tried to eat the bananas without peeling them. They refused, however, to try the smoked iguana which, as mentioned above, indeed may be one of their tabus. One Akuri, coming upon some near the mission hut, picked it up with his feet and while dexterously turning it about with his toes, solemnly repeated, "I do not eat your meat."

Many similarities to the superstitions of the Wayanas and Trios were noted. For instance, the extensive cutting of their arms in order to make them shoot straight; the hanging of certain bones in their huts and on their hammocks to insure the obtaining of meat on a future occasion; the refusal to eat certain meats; and the belief that birds and animals can verbally communicate with them.

Their animistic attitudes were particularly apparent in their spirit songs. Even on the first contact the girls sang almost all the time in low voices—almost as if they were singing to themselves. On the second contact, one male Akurijo said that just that morning a bird had flown through the camp crying, "Pigs are coming, pigs are coming." He seemed to be confidently awaiting their appearance while singing a song to attract them.

The Akurijos permitted the missionaries to photograph them during the September trip. Some examined the cameras, muttering, "Spirits," when they looked through the viewer and saw images of the mission group as well as their own tribesmen. They displayed curiosity mainly as the missionaries went about snapping pictures and only became alarmed when a few flash shots were taken. Then they said, "Hide that away, we are afraid of it." And the missionaries obliged.

The Akurijo form of salutation upon meeting appears to be hitting the chest with the fist and saying, "I am good," and hitting the back of the neck and saying, "I am one of your tribesmen." This is not a Trio custom and indeed may not be an Akuri custom; it was, however, something that developed during the course of the relationships occurring between the Trio and Wayana Indians in their first contact with this Akurijo band.

The Akuri prefer to call outsiders according to certain kinship terms. The missionaries felt that it was possibly equivalent to "brother." It is the term they use to address anyone else who comes to their band for the purpose of trading, and the visitor is then expected to address the Akurijo by the same term.

The object of this second contact was to place some of the most trustworthy Trio Indians with the Akuri tribe. They were to travel with the Akurijos for a few weeks for the purpose of gaining information about their language, territory, and customs, thus enabling the missionaries to better plan for further contact and a longer stay.

The Akuri had readily accepted the Indians in the mission group, for they are similar in appearance, carry similar weapons, and differ in dress only slightly. They squatted beside the Trios and Wayanas and freely looked through their hair for lice, combing and grooming them just as the Wayana and Trio Indians pick lice and groom each other in this age-old manner.

As for the rest of the party, so different in dress, language, etc., they were ignored by the Akurijos most of the time, though occasionally they pulled the hair on their arms to see if it hurt. This practice prompted one of the mission group to recall an account of a missionary who was killed completely out of curiosity. The Indian, thinking this unusual-looking creature might be a spirit, thrust his spear into the missionary's stomach to see whether or not it would go in. It did.

The Wayanas believe that the stone tools used by their forefathers had fallen out of the sky—a gift of one of their legendary spirits. It is possible that the Akurijos view the missionaries in much the same way, since they were the ones who gave them the axes and machetes that revolutionized their way of life and advanced their culture overnight from stone to steel.

So, although the Akuri were in good rapport with the Trios and Wayanas, they were not reluctant and indeed just a little less anxious to have the rest of the party leave than on the first contact. It was decided that five Trio men with food to last for one month would remain with the Akurijos and the rest of the party would quickly prepare to return downriver in order to conserve the food to be left for those remaining.

On the following morning, some of the men cached all the extra food on the other side of the river to prevent its being stolen once the Akurijos developed a taste for it.

Returning to the Akuri camp for the last time on this trip, the missionaries were told that most of them were out hunting, though the Trios felt they might have been hiding because of a lingering distrust and fear. As on the first contact, the Akuri were reluctant to have the missionaries remain with them for more than a few days. Therefore, the last day was spent in talking and trading for artifacts for the museums.

Of special interest were the tools made from rodents' teeth. A tooth imbedded in the end of a small hardwood stick is used for gouging and making holes in the seeds and monkey teeth that are strung for necklaces and aprons. The tooth is sharpened on the inside of the bark of a certain tree in the same manner one uses a sharpening stone. The long front teeth of the aimara (fish) are used for removing thorns and splinters from their feet.

After singing and prayer, the missionaries took their leave of the Akurijos. Some of them followed the mission group to the river and watched as they pulled away in their canoe. The Akuri had told the remaining five Trio men to get into the other boat and go too. However, when it was explained to them that they were going to stay on with the Akurijos, their only reply was, "Good." The missionaries' last glimpse of the camp as they paddled away was of the five Trios sitting on the bank and four Akurijos standing ankle deep in the water watching them depart.

It was understood that the Akurijos were planning to move farther west the next day and the Trios were prepared to travel with them. The Akurijos were expecting some others of their tribe, not yet seen by the mission group, to join up with them. Two other groups wandering about in the jungle between the Litani and Oelemari rivers were mentioned; one group of 4 men, the other of 10 men, with the numbers of women and children accompanying them left to speculation.

The Trios kept a short-wave radio with them and promised to be in touch with the mission group every Wednesday and Sunday morning at 6:30 A.M. They were instructed in the use of basic medicines, and promised to watch for any outbreak of serious illness. Although the Trios displayed real interest in living with the primitive Akuri for a time, they were aware of possible difficulties and promised to pull out of the area in the other canoe in the event of any hostility.

Only Trio men were placed with the Akurijos at this time. There was some discussion by the members of the mission expedition concerning the possibility of sexual problems among the Indians if the Trio wives were not brought along. But although the Trio women wanted to go, the Trio men felt they might not be able to keep up with the Akurijos—and other reasons were suspected as well. Consequently, only the strongest and most Christianized of the Trio men were selected to stay with the Akurijos. As it turned out, these men abstained but were seriously tempted by one Akurijo woman in particular who often made herself available. In fact, it was thought that one reason why the Trios left after three weeks was for fear of succumbing. Further discussion had it that abstention itself on the part of the Trios might have made them unusual to the Akurijos but, whatever the reactions were in this situation, it seemed to have in no way altered the very good contact between the two tribes.

In February 1969 a third expedition made contact with the same group of Akurijos for an eight-day period. However, at that time the Indians had separated into two temporary villages referred to by members of the expedition party as Atula's and Onore's groups, Atula and Onore being the chiefs of their respective villages. Although the Akurijo tribesmen in the first smaller village were friendly, those in the larger village, visited second, were hostile toward the missionaries and government officials though not to the Trio Indians. Therefore, when the expedition departed, three Trio Indians remained in each Akurijo village to teach and learn more about the Akuri people.

A fourth expedition made contact with a different group in late May and early June 1969. This group of 25 individuals, headed by Aparita (or Tupitas), was friendlier than either Atula's or Onore's groups. During this contact, another Akurijo chief of whom the West Indies Mission had heard, Araipuku, and his family came to Aparita's village. While some members of the expedition party went in search of Posoware, another Akurijo, who they had heard lived in relative isolation with his family, others remained with Aparita's group. The search party was unsuccessful and after four days rejoined the other members of the expedition party.

The party then proceeded to a place in the jungle known as the arrow cane area (so called because most the Akurijos make regular trips to the area to gather cane for making their arrows) where they agreed to locate a permanent village site for the Akurijos. The site was to consist of space for the village proper, a garden, and an air strip. This expedition was concluded on June 16.

In November 1969, another expedition set out to renew contact with Aparita, Araipuku, and their people with the intention of encouraging them to settle at the permanent village site. After only a single day of searching the expedition party came upon the Akurijos in one of their temporary villages.

The Indians were very friendly with expedition members, most of whom had been present at the May contact. The Akurijos were persuaded to accompany the expedition party to the permanent village site where a garden had only recently been planted. Although the

Indians were not impressed with the sprouting plants, they agreed to remain in the village as long as some of the expedition members remained with them.

On December 15 the missionaries, government representatives, and most of the Trio and Wayana Indians and Bush Negroes left the permanent village site. One Trio family remained with the Akurijos.

When the expedition returned in early March 1970, the Akurijos were still in the village and the planted fields were growing. On this trip a viral infection was introduced into the group. Both expedition members and Akurijos were stricken. Although there were no deaths and the missionaries nursed the Akurijos back to health, Aparita and some of his people were disillusioned and decided to leave the permanent village and return to their nomadic way of life. Araipuku and the remaining Akurijos agreed to stay in the village, and on March 20 the main expedition group departed, leaving the Akurijos in the care of two Trio families.

During previous contacts with the Akurijos, the missionaries had heard often of Posoware, who traveled alone with his family in the jungle north of the Oelemarie River. Six previous attempts had failed to locate Posoware. Now the largest expedition party of them all was assembled after extensive planning at the Wayana village of Big Arrow (Apetina). It consisted of one government representative, two Americans, four Bush Negro boatmen, one Akurijo boy, one Wayana Indian, and seventeen Trio Indians. Of the Trios, one was a young boy who dropped out after a few days and two were Trio women who joined the expedition in its final phase in order to accompany their husbands when they remained behind to teach the Akurijos.

On September 8, 1970, the entire expedition assembled at Big Arrow (Apetina). They left there at 1:00 P.M., traveling up the Pimba Creek to reach the fork at 3:00 P.M. Portage over the falls took another hour. It was 4:00 P.M. when they camped for the night at the left fork, known as the "black water creek," in an area filled with brilliant blue morpho butterflies. They found no signs of Akurijo huts or trails.

The young Akurijo boy traveling with the expedition seemed most eager to recontact his tribespeople. It was noticed he had three middle toes missing from his right foot and burns on his thighs. When questioned about this, he answered simply that his mother got mad at him and burned him. The cut marks on his body, he said, were made by his father.

On September 9 they made camp again at 10:20 A.M. and, since the rocks and rapids hindered their progress with the large canoes, they walked into the bush looking for Akurijo trails. The Bush Negroes had often reported seeing "wild Indians" in this area but no signs were found.

On September 10 they traveled seven kilometers, having left the canoes on the west side of Pimba Creek. The Trios killed several bush turkeys for dinner with their shotguns. At this, the Akurijo boy remarked that if Posoware heard the shots he would be afraid and flee deeper into the bush.

On September 11 the expedition party noted they made camp at two-ten and ten kilometers, beside a very small stream of water which permitted them "teacup baths." The next day disclosed a large clear stream just a few hundred meters from camp but no signs were seen of the Akurijos during the seven-kilometer search over the mountains. Sunday was devoted to rest and two services with the Indians.

On the fourteenth the Akurijo boy was not feeling well and it was decided to send him back to Apetina. The Trios made bark canoes to transport him down-creek and arrangements were made by radio to Apetina to have him picked up.

The expedition crossed a high mountain on the fifteenth and noted with amazement the incredible stamina of the Indians with their party, who continued to plod on over mountains and through swamps despite their very heavy back packs.

The sixteenth found them crossing another mountain and making camp number 8 where they observed the largest lightning bugs (fireflies) they had ever seen. They said they looked like large sparks from a fire and claimed their light increased ten times while in flight. After catching a few, the missionaries found they could actually use them in order to read their songbooks at night.

Camp was made on the seventeenth on a small creek at the foot of the largest mountain encountered on the trip. It was discovered later that Posoware regularly came to this creek to gather arrow cane.

On the eighteenth, after eleven days of traveling and not a single sign of Posoware or his tribespeople, the party made camp number 10 and decided to split up into three groups to search for trails, signs, etc. One group was to go south southeast, another due west and another south southeast on a 240-degree heading. At the last minute, this was changed to a 230-degree heading and despite the terrain this last group stuck to their course. After they were out about four kilometers, they came upon a beautiful valley and, as the missionaries remarked, "*Akurijo lon hele* [Akurijo ground this]," an Indian companion said, "*Enek* [Look]!" and there before them was an Akurijo trail. Since it was a "going and coming trail" (two-way) they pondered which way to go, or whether they should go at all as it was late and they might not be able to make it back to the base camp before dark. They decided to follow the trail east.

After approximately seven kilometers at a fast pace, they found evidence of a tree cut down for honey. A little farther up the trail they saw a tree that had been hacked upon with a stone ax—they agreed it was Posoware's!

Back to the base camp they went to report the good news. On the nineteenth the entire party left for the Akurijo trail. When they arrived at it they decided to go east and after traveling about eight to ten kilometers they came upon five Akurijo huts approximately two months old. These, they surmised, belonged to Akurijos of Epelite's group. A number of trails led in various directions from this camp site and in checking them out they found another hut made with stone axes about three hundred meters away. Again it was assumed this was Posoware's. They decided to camp there for the night and slept in the Akurijos' huts.

On the twentieth while a couple of the Indians of the expedition left to look for Posoware, the group moved camp three hundred meters northeast to a small clear creek as the stream on which they had camped was low and dirty.

Two other Indians went in search of Posoware in another direction on the twenty-first while the rest of the expedition party spent the day cutting trees and clearing an area for a food drop on the edge of a large bamboo patch. At 1:00 P.M. the Indians returned with Pileojana, the son of Posoware.

The news was that Posoware was alive and camped with another group (Amana's) on a stream they called *Late euku* (Late-bamboo). There had been some talk among the Akurijos that Posoware was dead—"*Posoware wateche, mato tetikai iweike* [Posoware is dead, his fire has gone out]."

Pileojana said that his father, Posoware, was not afraid. He and Amana had prepared the way obviously, as there had been sufficient contacts with these two previously to gain their confidence. Pileojana hung his hammock in one of their huts. Tied to his hammock was the shoulder bone of a turtle which he said was to keep him from getting fever. He said that Posoware had seen the airplane (*kura*). The possibility of at last seeing the elusive Posoware the next day was an exciting one.

Early the next morning, after Pileojana told of two other groups living in the bush, the expedition group packed up and moved back to the area next to the bamboo patch to await the Missionary Aviation Fellowship (M.A.F.) plane and the food drop. They set up a little battery-operated radio behind a tree (in order not to get bombed), stationed men behind the trees, and started the smoke signals going. The first two of the six drops went into the treacherous bamboo. The next four were right on target when directed from the radio from the ground. The cake and cookies were eaten immediately and the cassava meal and other items packed away and preparations were made to move out for Posoware's camp. They arrived there at 3:30 P.M., five kilometers from the drop area.

Only the women and children were at the Akurijo camp when the expedition party reached there. Some were from Epelite's group (a former contact) and the rest from Posoware's group. They said that Posoware was out gathering honey and hunting. The missionaries greeted the people and gave out a few gifts. They had heard of their approach and had surprisingly made two huts for them in welcome. After bathing, unpacking, and eating, all anxiously awaited the arrival of Posoware.

At about 6:00 P.M. Posoware and Amana walked into the camp carrying their bows and arrows. Posoware did not seem afraid or disturbed and his appearance surprised the expedition party more, it seemed, than they surprised him. He was young—they guessed he was perhaps thirty-three years old. Around his neck he wore the many strands of monkey-tooth necklaces which most Akurijos wear.

He was delighted with the gift of a small hatchet and said this would be his "tree-cutting thing" for gathering honey. He was just as pleased to receive another gift of a butcher knife. The Akurijos regularly used sharp bamboo to cut up their meat and also to cut their hair.

Although Posoware had never before seen a white man or a dark-skinned man and was at that moment seeing more people at one time than he had ever seen in his lifetime, he carried off the meeting very well, with a great deal of natural poise. At first he was slow and cautious but soon warmed up and was most co-operative in answering all the questions posed him.

Posoware's group consisted of just six people: Posoware, male, thirty-three years old, short, handsome, with a good physique, and in good health; his two wives: Ko, female, twenty-eight years old, healthy, mother of three living children, friendly, squinted noticeably with astigmatism in her left eye; Kurimaje, female, twenty years old, very sickly-looking, pale, weak, and just recovering from a miscarriage (her only other child, a boy, was killed accidentally by Posoware with a poison-tip arrow while hunting; she was the daughter of Irikine); two male children, ages four and two, and one female child, one year old. Posoware had one other boy, who died, he said, after getting caught in the rain and getting very cold.

The other Akurijos who had joined Posoware's group included Semokowala, Araipuku's former wife; Iripi, her son; Achane, her daughter (nursing); Amana; Maparu, wife of Amana; Meri, four-month-old baby of Maparu; Pileojana, son of Posoware; Napepejo, wife of Pileojana; and Kanaphe, mother of Pileojana.

The missionaries reported the obvious impact of a Christianized Trio, who had been living among the Akurijo group to which Pileojana and Amana belonged. They particularly were singing the gospel songs familiarly and the mission group noted that they heard the Akurijos singing these songs late into the night in contrast to a former expedition on which they heard them chant in a minor key from 7 P.M. until 2 A.M.

September 23 was devoted to a long interview with Posoware, beginning early in the morning. He told the mission group he had three names—Posoware, Warapoti, and Maripoti. His father, Kumatanepe, died at an old age downstream on the *Pilee euku* after a lifetime of roaming this area and the areas around the southern fork of the Oelemarie. Posoware's mother, known as Otoi, died somewhere between the food drop area and Posoware's camp, after the death of his father when Posoware was about ten years old. He was raised by his brother, Taripijana (Taripi is the meku monkey), who died at about forty, on the *Toli euku,* Posoware said, from a curse put on him by his tribespeople. When asked who cursed his brother, he replied that it was someone in the sky. The Akurijos believe that the dead go to the sky.

He revealed the fact that Ko, one of his wives, was his niece,

Taripijana's daughter. He also said that he had never seen a white man, a Bush Negro, or a canoe. He claimed to have seen the trail cuttings (made by sharp instruments) of others, which made him afraid. He said he had never eaten *kuwake* (cassava) nor did his mother ever tell him about cassava. (Nevertheless he ate it with enjoyment that day.) He knew nothing of pineapples, bananas, village life, fields, etc., nor had he heard of the former war between the Trios and a group of the Akurijos.

When the missionaries asked Posoware, "Why are you here on earth?" he replied, "Just to die." "Did your mother or father or brother tell you who made the trees?" asked the missionaries. "No," he said.

When they asked him where he was going now, he answered that he was on his way up the White Water Creek that runs off the Oelemarie to look for food and fiber of the iti palm with which to make hammocks.

His reply to "Why do you roam in such small groups?" and "Why don't you join up with some of the other Akurijos?" was, "That is just the way we are."

Posoware spoke of a group of people called Amiman (the little feet people) on the same trail that Irikine traveled. The missionaries had heard of them before, too, on a previous contact. He also said he had never seen houses, only trails, cut with sharp instruments.

He told them they got their stone axes on the Alapari Creek, an area later visited by one of the party and where many beautifully shaped stones were found. Posoware said they picked up stones already formed and sharpened them on any rock, demonstrating his remarks.

He told them he expected to go to the sky when he died though he believed Jolok, the chief evil spirit, lived in a village in the ground. What he feared most in the bush were spirits (*tijali*), jaguars or tigers, (*wili*), snakes (*matapi*), scorpions (*minete*), and tarantula spiders *(warikee)*. Actually many Akurijos have been killed by tigers.

Their name for the white people in the party, he said, was *Kaputawene eijamo,* meaning "sky ones" or "sky people." Obviously this made reference to the airplanes that drop food and transport the white people in and out of the area.

"Do you want to go to our village, Posoware?"

"Yes."

"In the airplane?"

"Yes, I am not afraid."

Posoware and his tribespeople were very friendly. They shared gifts with the expedition group and also their food after it had been cooked—the women in their little 7 by 9-inch seed aprons carrying palm fronds filled with cooked monkey to the expedition party made an unforgettable picture in the minds of their visitors.

At about 9:00 A.M. one morning during another interview with Posoware, someone was heard coming up the trail. Since all the members of both Posoware's group and the expedition were present and accounted for, there was much speculation about the approaching strangers. Suddenly Epelite and Kananaman walked into the camp clearing and Posoware leaped up, grabbing his bow and six arrows, and raced toward them. From that point on, things happened so fast that the mission group could barely tell what was going on. Pileojana grabbed Kananaman—they locked arms around their necks, shouting in each other's faces as they rocked back and forth as if dancing. Sometimes their action became so vigorous, they would fall to the ground. Later they exchanged partners and continued while the women in the camp calmly carried on their own activities, seemingly as uninterested and undisturbed as though their visitors dropped by every day. Actually they hadn't seen these people for five years. The greeting lasted for forty-five minutes to one hour and turned out to be a typical Akurijo "hello," though it had never been observed before by any of the expedition group.

One reason that Epelite and Kananaman came into Posoware's area was that they had heard the airplane and thought the mission party had come back to take them to the village. Secondly, they had come to report the death of Kananaman's wife (Amana's sister), a kind, friendly woman known to the missionaries from previous expeditions and of whom they had become very fond. When they asked what she died of, the answer was, "She ate *alawata* [red howler monkey], then her stomach hurt, then she died." Later they said she had dysentery.

The missionaries observed that Kananaman seemed to be taking his wife's death very well and the Trios commented that he had "opened his heart to the one in the sky" and that his wife had, too, before she died.

Eight days later one of the mission party personally visited the grave site located near an Akurijo camp on the Sarapi Creek. The grave site, which gave off a ghastly stench, was approximately thirty meters from their huts. It was a simple, crude, shallow grave and the

Akurijos described the burial in detail. They had scooped out a hole in the earth with sticks, then lined the grave with *mulu mulu ale* (thorn palm leaves). They placed her body on the leaves. She wore only her seed apron, plus one strand of white beads given her on an earlier contact. Leaves and dirt were placed upon her body, then thorn palm leaves and sticks were placed on the top of the grave as a seal or covering. She had been buried by Kananaman and Epelite—Epelite taking as his own all her monkey-tooth necklaces. They then made a *kata li* (back carrier) of leaf, put her hammock in it, and carried it off into the bush to discard it. She was not buried in her hammock in the manner of the Wayanas. Kanaphe, the old woman, after hearing of the death, waved an old woven basket to send the evil spirits away.

After the extended greeting, the hoarse and exhausted men settled back in their hammocks. Epelite and Kananaman said they had heard the airplane and hoped they could go in it to the Trio village on the western border of Surinam because "it is not good here, there are a lot of evil things and fever here." They said they wanted to go in the airplane and were not afraid. They remarked that they had left the women and children at another camp as they wanted to "go fast."

At that point two Trio Indians with the expedition party brought in six black spider monkeys—each day they shot meat for the Akurijos—and they were overjoyed at the sight of so much food and the prospect of "full stomachs" in the evening.

That evening Posoware appeared beside the hammock of one of the missionaries, holding up a cut and bloodied finger to be dressed. Although he displayed total unconcern for that procedure, he was fascinated by the missionary's flashlight, which he could only liken to fire without smoke or heat.

Early the next morning the interviews continued with the Trios as interpreters. Posoware told of his two trips over the mountain five years apart and of seeing Bush Negroes there. He said they did not shoot them but were prepared to if the Bush Negroes had shot at them first. He added that his tribespeople only shoot pigs. He also supplied some information (some of it conflicting) as to the whereabouts of Irikine—an Indian the missionaries had often heard about but had yet to encounter.

The Akurijos described Irikine as an old man with a deformed nose. They said it looked like a turtle's nose, bled from time to time, and that it had not been that way from birth. They volunteered the fact that it came from eating too much turtle meat, though a little would not be harmful. They said, deprecatingly, that Irikine only used sticks and reeds for arrows, without even tying feathers to them; that he had no bow wood and made his bows of inferior wood; that he did not draw his bow back very far; and that he ate only fish, turtles, and cayman, as he got no big game.

Epelite admitted eating iguana, but infrequently, because they do not travel the big rivers; and offered information about the location of sweet potatoes and fiber. Posoware told his interviewers laughingly that he used broken twigs on the trail not as trail markers or to help the women find their way but to get them out of his way. His answer to "What is bad (evil) to you?" was "Being angry, fierce, when others are about our women; to kill; to curse; to have evil or angry talk; one who doesn't answer back [common in Indian culture]; scolding [back and forth]; when evil or bad ones are among us; for a woman to have two husbands [but good for a man to have two wives]; to want revenge when someone takes your wife; to steal; to lie." As to stealing, their ideas vary—sometimes it is all right to steal from someone, but bad for someone to steal from you.

"Posoware, is it bad for one to take his brother's daughter for a wife?" This, in fact, Posoware had done. He replied, "It is a little good."

He also claimed that good to him meant taking these in proper relationship for a wife—especially two wives! He said that the women cut their legs with sharp bamboo so that they would hear well, obey well, and not be lazy. (He affirmed that the men cut their arms in order to shoot straight.) When asked at what age this was done and who did it, he said, "When they need it. When they are lazy, not obeying. They do it to themselves." At that moment Napepejo, the young wife of Pileojana, walked forward and displayed her cuts and said they had been made by Araipuku.

Posoware said he was a witch doctor and eagerly displayed his bag of charms, which contained the two old dead frogs he used to practice his witchcraft. He claimed his remedy for fever was to allow wasps to sting him, and the treatment for snakebite consisted of blowing smoke on the wound and/or rubbing ataki leaves on it. He also told the expedition party that his tribespeople did not wail after death but simply remained quiet and that the meats they did not eat were tiger, snake, and eagle.

It was late that night that Posoware's wife returned to the camp from the bush. She had left at 6:30 A.M. to search alone for sweet potatoes. It was pitch-black as she made her way back through the jungle with about ten potatoes about the size of quarters in her little woven basket.

On Sunday the twenty-seventh, Pileojana went hunting and returned with a wild pig. The old grandmother immediately set a piece of bark afire and, blowing upon the animal, rubbed the burning bark over the dead pig's body. Then, taking the ashes from the bark, she rubbed them on the head and back of her four-month-old grandchild. She openly explained this was to make the child walk at an early age so that it would not have to be carried by the mother on the trail. The Akurijo mothers, on the trail, regularly carry a heavy back pack full of possessions, often with a child riding atop it, and carrying a nursing child in their arms. The mother also often carries her child in a bark sling which is put around her head and then around the baby's buttocks. The baby learns very early to put his arms around her neck to hold on, thus freeing the mother's hands for chores.

The next day was spent observing the Akurijos at their various tasks. When Ko made a monkey-tooth necklace, she held the tooth in one hand and scraped the enamel off each side with a bone tool—an 8-inch stick with an *akuri* (agouti) tooth on the end. She then gouged or pierced a hole through it with an *aimala* (fish) tooth and strung the monkey teeth with a malajaime seed after each one on a piece of fiber.

Amana was observed making a stone ax. Working very quickly, he cut the wooden handle first, then gouged out the socket for the stone with an *akuri* (rodent) tooth.

On the twenty-ninth some of the group proceeded due west to try to locate the trails of Irikine. The search continued for three days with no positive results. On October 3 all expedition members met at the permanent village site. This area, known as the "field," had been cut and planted the previous year for the Akurijos and tended periodically since then by government officials, missionaries, and Indians in the hope they would one day make it their own village. It was now producing abundantly and here Kananaman saw more food in fifteen minutes than he had seen in his lifetime. He tasted cassava and sucked on sugar cane. He saw bananas, peppers, pumpkins, gourds, squash, sweet potatoes, coconut trees, and papaya. Later that day, on the expedition radio, he enthusiastically relayed to his people what he had seen.

The mission group spent the next several days working in the field and cutting a new large area with the hope of burning it in November and planting it in January.

By 4:30 A.M. on October 8 the expedition party was up and packing for their departure. Left behind at the field were two Trio families to teach the Akurijos the word of God, how to plant gardens, how to prepare these new foods, and to train them in the ways of village life. Left behind, too, was a radio for direct communication, simple medicines, and food.

At 6:50 A.M. the expedition left the field. By 11:00 A.M. they had reached the helicopter tower and by 5:00 P.M. had arrived at the Oelemari air strip. On October 9 they left the air strip at 7:00 A.M., reached the mouth of the Oelemari River at 12:30 P.M., came through the Gran Seela (rapids) without incident, and at 3:00 P.M. had returned to Kumakapan.

As of October 1970, the West Indies Mission had contacted five Akurijo groups, a total of 66 individuals. Of uncontacted groups, they knew about Irikine and his family (thought to be 5 individuals) and believed there were still other groups though they had no idea how many.

In summary, the following are some of the major observations recorded by the expedition members about the way of life of this nomadic people. Their diet consists mainly of *kwai* fruit, the thorn palm nut, meat, and honey—the constant search for the latter item providing the impetus for moving on from place to place. Their most important tool is the stone ax to which they refer as their "honey-gathering thing." In addition to the stone ax, the Akurijos possess several other primitive tools, notably bamboo knives, used for cutting hair and skinning animals, and tools made from rodents' teeth.

Monkey-tooth and seed necklaces are strung by men and women. Otherwise, the garb of the women is only a brief apron made of *malayaima* seeds, and that of the men, a narrow loincloth which they weave from *kuluwaitu* fibers—after making the "rack," the primitive loom, it takes about two days to complete these loincloths. The men also wear twine (plant-fiber) around their calves and ankles.

The bow and arrow is used for hunting. The bow is about 6⅓ feet long and is made with a boar's-tooth tool. Bowstrings are made from fibers of the Mauritia palm. Arrows of several types are used; some have long bamboo tips and others are made of bone. The Akurijos use curare poison and also carry poisoned arrow tips safely in a bamboo container slung around their necks.

Among the gifts presented to the Akurijos by the missionaries and government officials were steel axes, machetes, knives, and files.

The Surinam government has issued orders that all material Akurijo artifacts are to stay in the country, although they may remain in the possession of individuals. This is in accordance with their effort to establish close contact with the groups of the Akurijo tribe while protecting them from external exploitation and disease.

Although no studies of the tribe by professional anthropologists were known of at that time, the West Indies Mission undertook a program of educating the Akurijos to begin in 1971.

From early May 1971, the West Indies Mission has maintained two Trio Indian families with the two groups of Akurijos who still remain in the headwaters of the Litani River. These are the groups of Atula/Onore and Sokwiwi. The purpose is to provide a conditioning period for these Indians to help prepare them for eventual integration into the Trio tribe and villages. These are the last remaining groups of Akurijos thus far contacted who still remain in their own environment. These are also the ones who have been the least responsive to the encounters and who demonstrated hostility on several occasions.

The Trio families had gone in equipped with two short-wave radios, three batteries, a small generator, medicines, and food. Four other Trios accompanied them to help them get settled, cut an air drop area, and hollow out a canoe for use in case of emergency, or for their return tentatively set for September 1971. In the meantime, they were to be supplied by air drops every three weeks by the Missionary Aviation Fellowship.

Good progress was made though a few Akurijos remained aloof. Most of those who stole and were hostile in the early contacts became quite friendly. The one exception was Sokwiwi, who was still very easily aroused and angered.

This change in attitude was attributed in part to a notable shortage of natural foods at the time in the jungle. Jungle fruits, wild sweet potatoes, and other tubers were scarce and the Maurita palm (*kwai*) was not bearing fruit at all. It was for the purpose of gathering this fruit and making hammocks of the fibers in their palm leaves that the Akurijos had moved into this area for the rainy season.

The mission group supplemented the Akurijos' diet primarily with cassava bread and *kwak,* a kind of cereal made of cassava. The Akurijos became very fond of the cassava products and seemed quite content to stay at the drop area as long as the supply was plentiful. But when it dwindled they became very agitated and made it quite difficult for the Trios. They could not understand why food was brought to them rather than they being taken to the sources of the food, and kept demanding to be taken out by the Trios.

Within a few weeks it was reported that some of the Akurijos had colds and a number of them were infected with large undiagnosed sores. The sores are commonplace apparently to them, though they claimed to have contracted the colds in their recent contacts with Brazilians on the Matawale River in northern Brazil.

By the middle of June the Trios reported that the generator was no longer operating, the wife of Sokwiwi was seriously ill, and her newly born baby had died. Shortly thereafter, all communication with them was lost.

Permission was immediately requested and granted to make an emergency trip to the Akurijos to try to resolve these difficulties.

On June 24, 1971, the mission party, including one Trio Indian and four Wayanas, left the Lawa Mission Post for the headwaters of the Litani River in a ten-man canoe. Along the way they came upon the Trios' canoe, which had either been carried off by floodwaters or set adrift by the Akurijos. They later learned that the Trios had searched for it but because of high water and flooded banks were unable to find it. They were relieved and very pleased to have it towed back.

The Trios were in quite a difficult predicament. Their canoe was gone, the radios not working, food was low, and they had 19 irritable Akurijos on their hands. With the arrival of food, however, the Akurijos' attitude quickly changed.

It was soon discovered that the communication problem was due to dead batteries—the radios were in good working order. The generator had a frozen exhaust valve which was quickly repaired and as soon as the batteries were recharged, communications were restored.

The Akurijos gladly helped to carry the food and supplies over the trail from the canoe. There were 19 Akurijos with the Trios upon arrival. Seven more joined the group two days later and the following day the mission group carried one more in from the jungle where she had been left to die. Five Akurijos were missing and feared dead by their tribesmen.

Wives and husbands were shared by several in the group—often a woman was shared by father and son. Within the previous eighteen months one child had been killed by an anaconda and another by a jaguar. When Sokwiwi's wife became ill soon after delivering a baby girl, the infant quickly starved to death. The mother died shortly thereafter. Her symptoms were described as numerous large sores and swelling of the extremities and throat. Her young son clung to the corpse and refused to go with his father, saying his father would not give him food. The father told the Trios to take the boy and they were then caring for him.

The missionaries noted that several of the Akurijos had colds and coughed considerably during the night. Atula had an undiagnosed fever with pain in his back and knees. He appeared much thinner, had a poor appetite, and seemed very uncomfortable.

One of the babies had an enormous swelling on the side of its neck, about the size and shape of a small egg, though otherwise it seemed well. Superstitious about outsiders looking at their offspring, the mother covered its face and would not let it be treated or even looked at.

Onore had what appeared to be a fungus infection on his hands, feet, and lower legs with marked peeling of the skin.

Several of the Akurijos had large sores on their bodies. Some had them on their arms, legs, or backs and one teen-age boy had them all over his body, including his face. One child had them on his body in every stage of their development from a small pustule to an ulceration as large as the palm of a hand. Some were painful, others were not. The more purulent were accompanied by a deep, widespread cellulitis. Some sores which had either healed or were in a dormant stage appeared as large, raised, purplish welts, giving the skin a velvet appearance. Color slides of the sores were taken with a close-up lens so that they could be diagnosed and treated as soon as possible.

When the Akurijos who joined the group after the mission expedition arrived told of deserting a dying woman way back in the mountains, the missionaries set out to find her. Since in severe sickness and disability there is usually no hope of recovery, the Akurijos withhold food or desert the infirm ones, so that they will die quickly. They do this with seemingly little or no emotion; and in keeping with this philosophy, they also revile the aged when they can no longer provide for themselves and they become a burden to the rest of the group.

Along the trail, the missionaries met Sokwiwi, who told them the woman was still alive though dying and that he, too, had deserted her.

She was found in a small camp high in the mountains, huddled in her hammock over a smoldering fire—cold, hungry, and utterly without hope. She was emaciated, very weak, unable to stand, to stir the fire, or to get water. She also had a swollen, infected wound on her foot. The missionaries bathed and treated her and the Trios made a bark sling, packed her few possessions, and carried her out over the trail. The Trio men took turns carrying her piggyback, suspended from their heads in the bark sling.

The trail over the mountains and through vast palm swamps was very difficult but the men never complained or stumbled under their heavy load. Indeed, they were most cautious and considerate of the woman, using their hands to protect her legs from the tangle of vines, thorns, and saplings that impeded their progress. They were as pleased as the missionaries when they arrived back at the camp with the woman alive and much improved in spirit. The Akurijos, on the other hand, just laughed and said, "Why did you bring in those old bones?"

With the situation pretty well under control once again, the mission group prepared to leave. Several of the Akurijos at that point took their hammocks out to the canoe in the hope of being taken along.

All of the Akurijos except for Sokwiwi and Atula were desirous and even demanding to be brought out. They are few in number now and rapidly approaching extinction. Their food supply is limited, they lack tools, and incest is the rule rather than the exception in their tiny dwindling society and rivalries abound. There is no place for the weak, sick, handicapped, or aged and life is on a level little higher than the animals they hunt.

On the other hand, it is also obvious from all the above that if they are brought out the inroads made by civilization on their culture will probably doom them to a similar fate.

Part V
TABULATION OF EVENTS, 1968-73

All the reported events of five years, starting with the establishment of the Center for Short-Lived Phenomena in 1968, are listed chronologically within the four categories of Earth Science, Biological Science, Astrophysics and Anthropology.

The descriptive title, location and date of occurrence of each event are shown, together with the permanently assigned number under which it is documented in the reference files of the Center.

Earth Science Events

Event No.	*Name*	*Location*	*Date*
1–67	Deception Island Volcanic Eruption	South Shetland Islands	12/6/67
2–67	Metis Shoal Volcanic Eruption	Tonga Islands	12/11/67
1–68	Sicily Earthquake	Italy	1/15/68
6–68	Mount Mayon Volcanic Eruption	Philippines	4/20/68
7–68	Tokachi-Oki Earthquake	Japan	5/16/68
8–68	Inangahua Earthquake	New Zealand	5/23/68
9–68	Fernandina Volcanic Eruption	Galápagos Islands	6/11/68
11–68	Mount Etna Volcanic Eruption	Italy	6/16/68
12–68	Moyobamba Earthquake	Peru	6/19/68
17–68	Valparaiso Sea Surge	Chile	7/25/68
18–68	Kermadec Earthquake	New Zealand	7/25/68
20–68	Mount Arenal Volcanic Eruption	Costa Rica	7/29/68
21–68	Manila Earthquake	Philippines	8/1/68
22–68	Oaxaca Earthquake	Mexico	8/2/68
24–68	Molucca Passage Earthquake	Indonesia	8/10/68
30–68	Northern Celebes Earthquake	Indonesia	8/14/68
28–68	Kilauea Volcanic Eruption	Hawaii, U.S.A.	8/21/68
32–68	Iran Earthquake	Iran	8/31/68
33–68	Banua Wuhu Volcanic Eruption	Indonesia	9/6/68
44–68	Submarine Land Avalanche	New Guinea	9/16/68
35–68	Venezuela Earthquake	Venezuela	9/20/68
36–68	Guaratuba Substrata Collapse	Brazil	9/23/68
37–68	Raoul Island Earthquake	Kermadec Islands	9/26/68
41–68	Michigan Seiche	Michigan, U.S.A.	6/30/68
57–68	Turkey Earthquakes	Turkey	9/68
47–68	Taal Volcanic Activity	Philippines	10/2/68
48–68	Kilauea Volcanic Eruption	Hawaii, U.S.A.	10/7/68
43–68	West Australian Earthquake	Australia	10/14/68
50–68	New Guinea Earthquakes	New Guinea	9, 10/68
52–68	New Guinea Region Earthquake	New Guinea	10/23/68
53–68	Cerro Negro Volcanic Eruption	Nicaragua	10/27/68
56–68	Southern Illinois Earthquake	Illinois, U.S.A.	11/9/68
61–68	Mount Trident Volcanic Activity	Alaska	mid-11/68
63–68	Southwest Iceland Earthquake	Iceland	12/5/68
67–68	Santa Barbara Channel Earthquake Swarm	California, U.S.A.	12/28/68

Event No.	Name	Location	Date
1–69	Solomon Islands Earthquake	Solomon Islands	1/5/69
4–69	Mount Merapi Volcanic Eruption	Indonesia	1/8/69
8–69	Mindanao Earthquake	Philippines	1/30/69
12–69	Gunung Ija Volcanic Eruption	Indonesia	1/27/69
13–69	Telica Volcanic Activity	Nicaragua	2/11/69
16–69	Kilauea Volcanic Eruption	Hawaii, U.S.A.	2/22/69
17–69	Deception Island Volcanic Eruption	Antarctica	2/21/69
18–69	Macassar Strait Earthquake	Indonesia	2/23/69
19–69	Mount Ebu Lobo Volcanic Eruption	Indonesia	2/27/69
20–69	Portuguese Earthquake	Portugal	2/28/69
26–69	Minami-dake Volcano	Sakura-jima, Japan	3/8/69
29–69	Didicas Volcano	Philippines	3/27/69
30–69	Turkey Earthquake	Turkey	3/28/69
31–69	Halmahera Island Earthquake	Indonesia	3/27/69
32–69	Red Sea Earthquake	Red Sea Area	3/31/69
33–69	Ethiopian Rift Earthquake	Ethiopia	3/29/69
35–69	Norsälv Landslide	Sweden	4/12/69
36–69	Urkut Landslide	Hungary	3/14/69
41–69	Rincón de la Vieja Volcanic Activity	Costa Rica	4/22/69
43–69	Farallon de Pajaros Submarine Volcanic Eruption	Marianas Islands	3/11/69
44–69	Poas Volcanic Activity	Costa Rica	5/3/69
48–69	Amchitka Earthquake	Rat Island, Alaska	5/14/69
53–69	Mount Rainier Seismic Activity	Alaska, U.S.A.	1968–69
55–69	Salo Landslide	Finland	5/19/69
56–69	Shantung Tidal Wave	People's Republic of China	4/23/69
62–69	Ubinas Volcanic Activity	Peru	6/1/69
63–69	Becej Natural Gas Eruption	Yugoslavia	6/4/69
64–69	Mediterranean Sea Earthquake	Mediterranean	6/12/69
67–69	South China Flooding	People's Republic of China	4/23/69
70–69	Ruapehu Volcanic Eruption	New Zealand	6/22/69
71–69	Victorian Tremors	Australia	6/20/69
73–69	Castrovirreyna Sinking	Peru	6/25/69
79–69	Maghar Landslide	Israel	1, 2/69
82–69	Tolfa Tremors	Italy	7/2/69
86–69	Windward Passage Floating Island	Cuba	7/4/69
89–69	Corusqueira Landslide	Azores	6/25/69

Event No.	Name	Location	Date
92–69	Yellow Sea Earthquake	People's Republic of China	7/18/69
95–69	Peruvian Tremors	Peru	7/24/69
98–69	Molucca Passage Earthquake	Indonesia	8/5/69
101–69	Kurile Island Earthquake	Kurile Islands, U.S.S.R.	8/11/69
107–69	Tatabanya Mudflow	Hungary	8/15/69
110–69	Galápagos Islands Tremors	Galápagos Islands	8/30/69
111–69	Huehuetenango Landslide	Guatemala	8/28/69
114–69	Meije Rockfall	France	9/2/69
115–69	Amchitka Pass Earthquake	Aleutian Islands	9/12/69
116–69	Kiska Volcanic Eruption	Aleutian Islands	9/12/69
117–69	Davao "Land Rise"	Philippines	5/69
122–69	Cape Province Earthquake	South Africa	9/29/69
123–69	Santa Rosa Earthquake	California, U.S.A.	10/2/69
128–69	Huancayo Earthquake	Peru	10/1/69
130–69	Canlaon Volcanic Eruption	Philippines	10/11/69
131–69	Banja Luka Earthquake	Yugoslavia	10/26/69
132–69	California Channel Earthquakes	California, U.S.A.	10/22/69
133–69	Mount Taal Volcanic Eruption	Philippines	10/29/69
134–69	Kovachi Submarine Volcanic Eruption	Guadalcanal	10/30/69
135–69	Bezmianny Volcanic Eruption	U.S.S.R.	11/3/69
137–69	Southeast Iran Earthquake	Iran	11/7/69
139–69	Mentawai Trench Earthquake	Sumatra, Indonesia	11/21/69
140–69	Ostrov Karginskiy Earthquake	U.S.S.R.	11/22/69
141–69	Hawaiian Islands Storm Surge	Hawaii, U.S.A.	12/2/69
145–69	Leeward Islands Earthquake	Guadeloupe	12/25/69
1–70	Kunming Earthquake	People's Republic of China	1/4/70
3–70	Pacaya Volcanic Eruption	Guatemala	1/7/70
4–70	Davao Earthquake	Philippines	1/10/70
6–70	Mount Ulawun Volcanic Eruption	New Guinea	1/16/70
7–70	Kermadec Islands Earthquake	South Pacific	1/20/70
9–70	Mount Lokon Volcanic Eruption	Celebes, Indonesia	11/27/69
10–70	Myojinsho Volcanic Explosion	South Izu Islands, Japan	1/29/70
14–70	Kilauea Volcanic Activity	Hawaii, U.S.A.	1970
16–70	Tonga Submarine Volcanic Activity	South Pacific	1/3/70

Event No.	Name	Location	Date
17–70	Imessi-ile Boulder Slide	Nigeria	9/1/69
20–70	Pozzuoli Uplift	Naples, Italy	3/70
25–70	Calingiri Earthquake	Western Australia	3/10/70
28–70	Gujarat Earthquake	India	3/23/70
29–70	Gediz (Western Turkey) Earthquake	Turkey	3/28/70
31–70	Southern Mindanao Earthquake	Philippines	3/30/70
32–70	Chaparraspique Volcanic Activity	El Salvador	3/30/70
33–70	Luzon Earthquake	Philippines	4/7/70
34–70	Gulf of Alaska Earthquakes	Alaska, U.S.A.	4/16/70
35–70	Mount Malinao Volcanic Activity	Philippines	4/24/79
37–70	Guatemala Basin Earthquake	Offshore Central America	4/29/70
38–70	Masaya Volcanic Activity	Nicaragua	4/4; 5/70
39–70	Mount Aso Volcanic Eruption	Kyodo, Japan	4/21/70
43–70	Hekla Volcanic Eruption	Iceland	4/21/70
44–70	Guayas Province Flooding	Ecuador	5/5/70
47–70	Karimsky Volcanic Eruption	Kamchatka, U.S.S.R.	5/12/70
48–70	Southeastern Europe Flooding	Rumania	5/70
49–70	Huaraz Earthquake	Peru	5/31/70
52–70	Sino-Soviet Border Earthquake	U.S.S.R.	6/5/70
53–70	Suwanose-shima Volcanic Explosion	Ryukyu Islands, Japan	5/28/70
58–70	Queen Charlotte Islands Earthquake	British Columbia, Canada	6/24/70
59–70	Cheldag Mud Volcano Activity	Baku, U.S.S.R.	6/70
60–70	Sakura-jima Volcanic Explosion	Kyushu, Japan	5, 6/70
62–70	Damghan Earthquake	Iran	7/30/70
63–70	Amazonas Earthquake	Colombia	7/31/70
68–70	New Hebrides Islands Earthquake	New Hebrides Islands	8/11/70
70–70	Deception Island Volcanic Activity	South Shetland Islands	8/13/70
73–70	Rincón de la Vieja Volcanic Activity	Costa Rica	8/26/70
75–70	Telica Volcanic Activity	Nicaragua	8/18/70
77–70	Lake Linda Drainage	Juneau, Alaska, U.S.A.	8/70
82–70	Dunaföldvár Landslide	Hungary	9/15/70
85–70	Akita-Koma-ga-take Volcanic Eruption	Japan	9/18/70
86–70	Wolenchiti Fracturing	Ethiopia	8/70

Event No.	Name	Location	Date
89–70	Beerenberg Volcanic Eruption	Jan Mayen Island	9/25/70
95–70	Alcedo Volcanic Activity	Galápagos Islands	10/26/70
97–70	Madang Earthquake	New Guinea	10/31/70
100–70	Bay of Bengal Tidal Waves	East Pakistan	11/13/70
103–70	Typhoon Patsy	Manila, Philippines	11/20/70
108–70	Caribbean Submarine Volcanic Activity (Mud Volcano)	Colombia Coast	12/2/70
109–70	Peru-Ecuador Border Earthquake	Peru-Ecuador Border	12/9/70
113–70	Mount Taal Volcanic Activity	Philippines	12/29/70
1–71	West Irian Earthquake	New Guinea	1/10/71
3–71	West Malaysia Flooding	West Malaysia	12/30/70
4–71	Santa Fe Fumarole Activity	Galápagos Islands	11/1/70
14–71	Kepulauan Mentawai Earthquake	Indonesia	2/4/71
15–71	Cerro Negro Volcanic Eruption	Nicaragua	2/3/71
17–71	San Fernando Earthquake	U.S.A.	2/9/71
19–71	Tuscania Earthquake	Tuscania, Italy	2/6/71
20–71	Argentine Antarctic Earthquake	Antarctica	2/8/71
21–71	Nyiragongo Volcanic Activity	Zaire	12/24/70
22–17	Erta'ale Volcanic Activity	Ethiopia	12/70 cont.
23–71	Karua Submarine Volcanic Eruption	New Hebrides Islands	2/22/71
27–71	Chungar Avalanche	Peru	3/18/71
31–71	Nyamuragira Volcanic Activity	Zaire	3/24/71
33–71	Mount Etna Volcanic Activity	Italy	4/5/71
34–71	Stromboli Volcanic Activity	Italy	4/3/71
36–71	Ruapehu Volcanic Activity	New Zealand	3–5/71
41–71	Adak Island Earthquake	Alaska, U.S.A.	5/2/71
42–71	Galápagos Earthquake Swarm	Galápagos Islands	4/5/71
43–71	Azufral de Tuquerres Seismic Activity	Colombia	4/26/71
44–71	Burdur Earthquake	Western Turkey	5/12/71
46–71	St. Jean Vianney Landslip	Canada	5/4/71
50–71	Turkey Eastern Province Earthquake	Turkey	5/22/71
55–71	Mojjo Fracturing	Ethiopia	5/25/71
59–71	Milan Severe Hailstorm	Italy	6/14/71
60–71	Salvador Floods	Brazil	4/26–30/71
61–71	Chile-Argentina Border Earthquake	Chile-Argentina	6/17/71
66–71	Valparaiso Earthquake	Chile	7/9/71

Event No.	Name	Location	Date
69–71	Solomon Islands Earthquake	Solomon Islands	7/14/71
71–71	New Britain Earthquake	New Guinea	7/26/71
72–71	Morona Santiago Earthquake	Ecuador	7/27/71
73–71	White Island Volcanic Activity	New Zealand	7/19/71
75–71	Hidaka Earthquake	Japan	8/2/71
76–71	Romanche Gap Earthquake	Mid-Atlantic Ocean	8/5/71
80–71	Mount Hudson Volcanic Activity	Chile	8/12/71
81–71	Kilauea Volcanic Eruption	Hawaii, U.S.A.	8/14/71
85–71	Fuego Volcanic Activity	Guatemala	9/14/71
87–71	Tinakula Volcanic Eruption	Santa Cruz Islands, South Pacific	9/71
88–71	Bougainville Earthquake	Solomon Islands	10/4/71
90–71	La Palma Volcanic Activity	Canary Islands	10/26/71
91–71	Espiritu Santo Earthquake	New Hebrides Islands	10/27/71
92–71	Soufrière Volcanic Activity	St. Vincent, West Indies	10/31/71
93–71	Utupua Earthquake	Santa Cruz Islands, South Pacific	11/21/71
95–71	Villarrica Volcanic Eruption	Chile	10/29/71
96–71	Kamchatsky Proliv Earthquake	U.S.S.R	12/15/71
99–71	San Cristóbal Volcanic Activity	Nicaragua	5/3/71
100–71	Telica Volcanic Activity	Nicaragua	12/1/71
2–72	Izu-O-shima Earthquake Swarm	Japan	1/14/72
3–72	Taiwan Earthquake	Taiwan, China	1/26/72
5–72	Torres Island Earthquake	New Hebrides Islands	1/23/72
9–72	Kilauea Volcanic Eruption	Hawaii, U.S.A.	2/5/72
11–72	Pacaya Volcanic Eruption	Guatemala	2/2/72
13–72	Hachijo-jima Earthquake	Japan	2/29/72
14–72	Nyiragongo Volcanic Eruption	Zaire	1/29/72
15–72	Puyehue Volcanic Activity	Chile	3/1/72
16–72	Erta Ale Volcanic Activity	Ethiopia	2/72
18–72	Recice Landslides	Czechoslovakia	1/72
20–72	Grímsvötn Jökulhlaup	Iceland	3/72
22–72	Jahrom Earthquake	Iran	4/10/72
27–72	Mount Semeru Volcanic Activity	East Java, Indonesia	4/72
29–72	Hua-Lien Earthquake	Taiwan, China	4/24/72
30–72	Manila Earthquake	Philippines	4/25/72
33–72	Kurile Islands, Submarine Volcanic Activity	Kurile Islands, U.S.S.R.	4/29/72

Event No.	Name	Location	Date
36–72	Sangihe Earthquake	Archipelago, Indonesia	6/11/72
37–72	Hong Kong Landslides	Hong Kong	6/18/72
38–72	Alaid Volcanic Eruption	Kurile Islands, U.S.S.R.	6/72
49–72	Sitka Earthquake	Alaska, U.S.A.	7/30/72
52–72	New Ireland Earthquake	New Guinea	8/17/72
54–72	Vanikoro Island Earthquake	Santa Cruz Islands, South Pacific	9/4/72
57–72	Karthala Volcanic Eruption	Comoro Islands, Indian Ocean	9/8/72
59–72	Banda Sea Earthquake	Indonesia	9/24/72
60–72	Sakura-jima Volcanic Eruption	Kyushu, Japan	8/72
61–72	Krakatoa Volcanic Activity	Indonesia	6/26/72
62–72	Bad Schandau Earthflow	Dresden, Germany	9/10/72
63–72	Poas Volcanic Eruption	Costa Rica	6/21/72
67–72	Mount Merapi Volcanic Activity	Indonesia	10/6/72
69–72	West Transdanubian Floods	Hungary	7/72
71–72	Piton de la Fournaise Volcanic Eruption	Indian Ocean	6/18/72
74–72	Tanna Island Earthquake	New Hebrides Islands	11/2/72
77–72	Somalia Cyclone	Somalia, Aden, and Afars-Issas Territory	10/24/72
78–72	Big Sur Mud Slides	California, U.S.A.	10/72
79–72	Michigan-Ohio Shore Line Flooding	U.S.A.	11/14/72
80–72	Queensland-New South Wales Floods	Australia	10/72
82–72	Ritter Island Eruptions	New Guinea	10/8/72
83–72	Western U. S. Coastal Wave Action	California, U.S.A.	11/25/72
87–72	Funafuti Storm Ridge Development	South Pacific	10/20/72
88–72	Eastern Mindanao Earthquake	Philippines	12/2/72
89–72	Japan Trench Earthquake	Japan	12/4/72
94–72	Acatenango Volcanic Activity	Guatemala	11/72
96–72	Managua Earthquake	Nicaragua	12/23/72

Biological Science Events

Event No.	*Name*	*Location*	*Date*
3–68	Ocean Eagle Spill	San Juan, Puerto Rico	3/3/68
39–68	Louisiana Fish Kill	Louisiana, U.S.A.	6/13/68
10–68	World Glory Spill	South Africa	6/14/68
14–68	California Fish Kill	California, U.S.A.	6/24/68
40–68	Vermont Windstorm	Vermont, U.S.A.	6/27/68
13–68	Florida Fish Kill	Florida, U.S.A.	6/27/68
16–68	Newcastle Mussel Poisoning	England	7/68
23–68	Sudan Locust Swarm	Republic of Sudan	8/6/68
25–68	Virginia Fish Kill	Virginia, U.S.A.	8/12/68
27–68	Chilean Drought	Chile	3–8/68
42–68	Accabonac Harbor Fish Kill	New York, U.S.A.	8/29/68
34–68	Appalachian Squirrel Migration	North Carolina, Tennessee, U.S.A.	9/10/68
45–68	Mississippi Flyway Duck Migration	Mississippi Valley, U.S.A.	9/10/68
46–68	Massachusetts Maple Tree Deaths	Massachusetts, U.S.A.	9/68
49–68	California Seal Colonization	San Miguel Island	10/14/68
55–68	Allegheny River Oil Spill/Fish Kill	Pennsylvania, U.S.A.	10/29/68
58–68	Hawaiian Monk Seal Disappearance	Midway Atoll	11/68
60–68	*Hess Hustler* Oil Spill	Delaware, U.S.A.	11/13/68
66–68	South Sargasso Sea Marine Fouling	Sargasso Sea	12/5/68
64–68	*Witwater* Oil Spill	Panama	12/13/68
68–68	Wakulla Oil Slick	Florida, U.S.A.	12/30/68
9–69	Santa Barbara Oil Spill	California, U.S.A.	1/28/69
10–69	St. Mary's Duck Kill	Maryland, U.S.A.	1/23/69
15–69	Anchorage Moose Migration	Alaska, U.S.A.	2/69
22–69	Peruvian Ant Irruption	Peru	3/1/69

Event No.	Name	Location	Date
23–69	Tecolutla Sea Animal Discovery	Mexico	3/6/69
24–69	Cook Inlet Oil Spill	Alaska, U.S.A.	3/4/69
25–69	Louisiana Oil Spill	Louisiana, U.S.A.	3/16/69
27–69	Chicago River Oil Spill	Illinois, U.S.A.	3/21/69
28–69	Iceland Polar Bear Kill	Iceland	1/22/69
38–69	New Jersey Fish/Crustacean	New Jersey, U.S.A.	9, 10/68
39–69	Dutch Coast Oil Spill	Netherlands	2/20/69
45–69	Great Barrier Reef Starfish Plague	Australia	1968–69
46–69	Samar Spontaneous Soil Burn	Philippine Islands	4/20/69
47–69	Sandy Hook Fish Kill	New Jersey, U.S.A.	5/11/69
49–69	*Hamilton Trader* Oil Spill	England	4/30/69
51–69	Lake Michigan Salmon High DDT Residue Discovery	Michigan, U.S.A.	3/69
54–69	Piscataqua River Oil Spill	New Hampshire, U.S.A.	5/69
57–69	Junín Lagoon Marine Pollution	Peru	5/12/69
60–69	Weymouth Oil Spill	Massachusetts, U.S.A.	6/7/69
65–69	Raleigh Bay Bird Kill	North Carolina, U.S.A.	6/13/69
68–69	Alma Oil Spill	Wisconsin, U.S.A.	6/16/69
69–69	Rhine River Fish Kill	Rhine River Europe	6/19/69
74–69	Cook Inlet Oil Spill	Alaska, U.S.A.	6/23/69
76–69	Chili River Contamination	Peru	6/14/69
78–69	Toledo Cattle Kill	Spain	1968–69
81–69	Newport Clam Irruption	Rhode Island, U.S.A.	6/69
83–69	Leine River Pollution	Germany	7/3/69
84–69	Chirimachas Advance	Peru	1968–69
85–69	Sooty Tern Hatch Failure	Florida, U.S.A.	5, 6/69
87–69	Massachusetts Bird Kill	Massachusetts, U.S.A.	6/29/69
88–69	Amazon River Porpoise Disappearance	Peru	1968–69
90–69	Japanese Bamboo Kill	Japan	1969
91–69	United Kingdom Bird Kill	United Kingdom	7/6/69
93–69	Charles River Fish Kill	Massachusetts, U.S.A.	7/18/69
97–69	Southern Lake Michigan Alewife Mortality	Michigan, U.S.A.	7–8/69
100–69	Northeastern Gypsy Moth Infestation	U.S.A.	7–8/69

Event No.	*Name*	*Location*	*Date*
103–69	South Africa: Oiling of Penguins	Republic of South Africa	8/10/69
105–69	Waterfowl Kill—Botulism	U.S.A.	8/10/69
108–69	*Gironde* Oil Spill	St. Brieuc, France	8/19/69
109–69	Eastern U. S. High Air Pollution Potential	U.S.A.	8/69
112–69	Panama Moth Migration	Panama	8/13/69
113–69	European Migration Bird Scarcity	Europe	1969
118–69	West Falmouth Oil Spill	Massachusetts, U.S.A.	9/16/69
119–69	Trinidad Butterfly Invasion	Trinidad, West Indies	9/69
121–69	Miami Snail Infestation	Florida, U.S.A.	9/69
127–69	Great Britain Bird Mortality	United Kingdom	10/69
129–69	St. Louis Spider Invasion	Missouri, U.S.A.	10/8/69
136–69	Westport Snake Mortality	Canada	10/13/69
138–69	Golden-crowned Kinglet Mortality	Canada	10/23/69
142–69	New Jersey-New York Crossbill Irruption	New Jersey, New York, U.S.A.	11/69
143–69	Florida Spanish Moss Mortality	Florida, U.S.A.	1969
144–69	Great Britain Seal Mortality	United Kingdom	11–12/69
5–70	Florida Beached Whale Mortality	Florida, U.S.A.	1/11/70
11–70	Tunisian Nematode Infestation	Tunisia	1969–70
12–70	British Isles Sea Bird Mortality	United Kingdom	12/69–1/70
13–70	Martha's Vineyard Bird Mortality	Massachusetts, U.S.A.	2/70
15–70	*Arrow* Oil Spill	Nova Scotia, Canada	2/4/70
18–70	*Delian Apollon* Oil Spill	Florida, U.S.A.	2/13/70
19–70	Southwest Sweden Snow Pollution	Sweden	12/69–1/70
21–70	Victorian Mouse Plague	New South Wales, Australia	3/70
22–70	Breton Sound Oil Spill	Louisiana, U.S.A.	3/10/70
23–70	Scotland Neck Blackbird Irruption	North Carolina, U.S.A.	1969–70
24–70	*Oceanic Grandeur* Oil Spill	Torres Strait, Australia	3/3/70
27–70	Chesapeake Bay Sea Duck Mortality	Maryland, Virginia, U.S.A.	2/70
36–70	Alaska Peninsula Oil Spill	Alaska, U.S.A.	4/25/70
40–70	Lake St. Clair and Lake Erie Mercury Contamination	U.S.A. and Canada	1970
41–70	Tarut Bay Oil Spill	Saudi Arabia	4/20/70
42–70	*Othello* Oil Spill	Trälhavet Bay, Sweden	3/20/70

Event No.	Name	Location	Date
45–70	Crooked Creek Fish Kill—Pesticides	Missouri, U.S.A.	5/5/70
46–70	*Polycommander* Oil Spill	Spain	5/5/70
51–70	Älvsbyn Reindeer Mortality	Sweden	4–5/70
54–70	Moroccan Grain Crop Destruction	Morocco	1970
55–70	Imminent Kouprey Extinction	Cambodia	Summer 1970
57–70	Brazilian Coffee Rust	Brazil	1970
66–70	Konert Spring Gas Leak	Missouri, U.S.A.	7–8/70
71–70	Lake Erie Mud Puppy Mortality	U.S.A. and Canada	8/19/70
72–70	Northeastern U. S. Caterpillar Infestation	U.S.A.	1968 cont.
74–70	U. S. Southern Corn Leaf Blight	U.S.A.	1970
76–70	Escambia Bay Menhaden Kill	Florida, U.S.A.	9/3/70
78–70	Studley Pond Fish Kill	Massachusetts,	8/70
80–70	Virginia Beach Fish Kill	Virginia, U.S.A.	8/21/70
84–70	Tucson Snail Infestation	Arizona, U.S.A.	9/24/70
87–70	Holston River Mercury Contamination	Virginia, U.S.A.	9/22/70
88–70	Monarch Butterfly Migration	U.S.A., Canada, Mexico	8–10/70
90–70	Sea Bird Egg Contamination	California, U.S.A.	1970 cont.
92–70	Southern Gulf of Mexico Fish Kill	Gulf of Mexico	10/22/70
93–70	Japanese Tanker Gasoline Spill	Japanese Coast	10/16/70
96–70	Isle of Wight Oil Spill	England	10/23/70
98–70	Rio Solimões Fish Kill	Brazil	9/16/70
99–70	Malaysian Frog War	Malaysia	11/8/70
101–70	Schuylkill Oil Spill	Pennsylvania, U.S.A.	11/13/70
102–70	San Bernardino Brush Fires	California, U.S.A.	11/13–17/70
104–70	Panama Pollution Potential	Panama	11/28/70
105–70	Louisiana Oil Drill Platform Fire	Louisiana, U.S.A.	12/1/70
106–70	Florida Keys Oil Slick	Florida, U.S.A.	12/1/70
107–70	Jacksonville Oil Slick	Florida, U.S.A.	12/2/70
111–70	Formentera White-tailed Rat Invasion	Spain	12/12/70
112–70	Litoral de Pisco Fish Kill	Pisco, Peru	12/18/70
2–71	Amer Oil Spill	Netherlands	12/12/70
5–71	San Clemente Island Beached Whales	U.S.A.	1/8/71
6–71	San Francisco Bay Oil Spill	U.S.A.	1/18/71
8–71	Tanzania Army Worm Outbreak	Tanzania	11/70 cont.
9–71	Long Island Sound Oil Spill	U.S.A.	1/24/71
10–71	Garrison, Utah, Sheep Kill	U.S.A.	1/20/71
18–71	Point Noir Sea Swallow Arrival	Zaire	11–12/70

Event No.	Name	Location	Date
24–71	Wafra Oil Spill (Stilbaai)	South Africa	2/27/71
25–71	Florida Aphid Outbreak	Florida, U.S.A.	Winter 1970, 1971
28–71	Soviet Oil Pipe Leak	U.S.S.R.	3/21/71
29–71	*Texaco Oklahoma* Oil Slick	U.S.A.	3/27/71
30–71	Thuntank 6 Oil Spill	South Wales, United Kingdom	3/14/71
37–71	*Paracas-Texaco Caribbean-Brandenburg* Oil Spill	United Kingdom	1/11/71
38–71	*Noctiluca miliaris* Plankton Bloom	Spain	4/18/71
39–71	Cape Hatteras-Rehoboth Beach Starfish Kill	North Carolina and Maryland, U.S.A.	4/23–25/71
40–71	*Panther* Oil Spill	English Channel	3/30/71
45–71	York River Oil Spill	Virginia, U.S.A.	5/5/71
47–71	Northeastern Minnesota Forest Fire	Minnesota, U.S.A.	5/14/71
48–71	South Florida Everglades Drought	Florida, U.S.A.	4–5/71
49–71	Anacortes Oil Spill	Washington, U.S.A.	4/26/71
52–71	Honduras Forest Fires and Smoke Pollution	Honduras	4–5/71
53–71	White Fly Plague	Spain	Summer 1969
54–71	Saiga Antelope Wheat Invasion	U.S.S.R.	1971
56–71	Sule Skerry Islet Bird Kill	Scotland	4/26/71
57–71	Shetland Islands Bird Kill	Scotland	5/30/71
62–71	Liverpool Naphtha Escape	United Kingdom	6/12, 13/71
63–71	Espolla Lagoon Rare Crustacean Disease	Spain	6/3/71
65–71	Arabian (Persian) Gulf Fish Kill	Saudi Arabia	6/22/71
68–71	St. Petersburg Red Tide	Florida, U.S.A.	6/71
70–71	*Towle* Oil Spill	Bayonne, N.J., U.S.A.	7/14/71
74–71	*Squilla armata* Population Increase	South Africa	5, 6/71
77–71	Lee's River Fish Kill	Massachusetts, U.S.A.	8/5/71
78–71	Green Pond Fish Kill	Massachusetts, U.S.A.	8/5/71
82–71	San Clemente Oil Spill	California, U.S.A.	8/20/71
83–71	Northwest Atlantic Puffin Decline	Atlantic Ocean	1971
86–71	Monarch Butterfly Peak	North America	Fall 1971
89–71	Amoco Oil Platform Fire	Gulf of Mexico	10/18/71
94–71	*Juliana* Oil Spill	Niigata, Japan	11/30/71

Event No.	Name	Location	Date
97–71	Peace River Phosphate Spill	Florida, U.S.A.	12/3/71
98–71	Laban Island Oil Spill	Persian Gulf	12/2/71
1–72	West Fayu Oil Spill	Caroline Islands	12/30/71
4–72	*Germania* Chemical Pollution	English Channel	1/72
6–72	Homestead Bird Kill	Florida, U.S.A.	1/28/72
7–72	Oyster Creek Fish Kill	New Jersey, U.S.A.	2/7/72
8–72	East Florida Coast Common Loon Mortality	Florida, U.S.A.	1/25/72
10–72	Firth of Forth Sea Duck Kill	Scotland	2/4/72
12–72	Melbourne Beached Whales	Australia	2/3/72
17–72	Gray Fox Epizootic Canine Distemper	New York, U.S.A.	2/18/72
19–72	*FL Hayes* Oil Spill	Connecticut, U.S.A.	3/21/72
21–72	*Dewdale* Oil Spill	Scotland	3/20/72
23–72	Tasmanian Coast Wave Damage	Australia	3/25/72
28–72	Chino Farm Goose Die-off	Maryland, U.S.A.	3/11/72
31–72	*Tien Chee* Oil Spill	Uruguay	5/11/72
34–72	Charleston Nitric Acid Release	Illinois, U.S.A.	5/29/72
35–72	Canadian Lynx Imminent Invasion	Minnesota, U.S.A.	5/72
39–72	Britain Gannet and Shag Mortality	Coast of Britain	2/72
40–72	Queensland Mouse Plague	Australia	1972
41–72	San Francisco Dutch Duck Plague Outbreak	California, U.S.A.	4/72
42–72	Transandean Oil Pipeline Leak	Ecuador	6/30/72
43–72	Spain Olive Fly Plague	Spain	6/72
44–72	Possible Capercaillie Extermination	Spain	1972
45–72	Pennsylvania Martin Mortality	Pennsylvania, U.S.A.	6/72
46–72	*Tamano* Oil Spill	Maine, U.S.A.	7/22/72
48–72	Manzini Oxidation Ponds Fish Kill	Swaziland	7/21/72
53–72	Bear Fire	California, U.S.A.	8/22/72
55–72	Essex County Duck and Gull Kill	Massachusetts, U.S.A.	9/14/72
55–72	Massachusetts Red Tide	Massachusetts, U.S.A.	9/14/72
56–72	*República de Colombia* Oil Spill	Off Cape Hatteras, U.S.A.	9/14/72

Event No.	Name	Location	Date
58–72	*Oswego Guardian* Oil Spill	South Africa	8/21/72
64–72	Salem Harbor Oil Tank Spill	Massachusetts, U.S.A.	9/27/72
65–72	Meuse Valley Toxic SO_2 Emission	Belgium	9/5/72
66–72	Miami Palms Lethal Yellowing	Florida, U.S.A.	9/71
68–72	Western Pacific Smoke and Haze Incidence	Western Pacific Ocean	9/5/72
70–72	Texas-New Mexico Oil Pipeline Spill	Northwest of Arizona, U.S.A.	10/10/72
72–72	*Genimar* Oil Spill	English Channel	10/1/72
73–72	Ocean 80 Barge Explosion and Spill	New Jersey, U.S.A.	10/25/72
75–72	Pah River Fire	Alaska, U.S.A.	7/8/72
76–72	Brazilian Bee Infestation	South America	1956 cont.
81–72	Paraguayan Leaf-cutting Ant Plague	Paraguay	1972
84–72	Arthur Kill Oil Spill	New Jersey, U.S.A.	11/28/72
85–72	North Atlantic Ocean Oil Spill	North Atlantic Ocean	11/27/72
86–72	Cimarron Canyon Diesel Fuel Spill	New Mexico, U.S.A.	10/24/72
90–72	Southwest Pass Well Blowout	Gulf of Mexico, U.S.A.	12/3/72
91–72	*Vita* Oil Spill	New York, U.S.A.	12/5/72
92–72	Goshawk High Migration Number	Minnesota, U.S.A.	1972
93–72	North Sea Oil Spill and Bird Kill	Denmark	12/72
95–72	North Rhine-Westphalia Oil Pipeline Break	West Germany	12/5/72
97–72	Bouchard No. 40 Oil Spill	Long Island, New York, U.S.A.	12/25/72
98–72	Rhine River Ammonia Spill	Germany	12/22/72
99–72	Connecticut Oil Spill	Connecticut, U.S.A.	12/26/72

Astrophysical Events

Event No.	*Name*	*Location*	*Date*
2–68	Polo Fireball	Illinois, U.S.A.	2/26/68
4–68	Dayton Fireball	Ohio, U.S.A.	3/3/68
5–68	Veracruz Fireball	Mexico	3/27/68
38–68	Schenectady Meteorite	New York, U.S.A.	4/12/68
15–68	Huntingdon Fireball	Pennsylvania, U.S.A.	7/2/68
19–68	Omaha Fireball	Iowa, U.S.A.	7/28/68
26–68	Comodoro Rivadavia Fireball	Argentina	8/17/68
29–68	Albany Fireball	New York, U.S.A.	8/27/68
31–68	Grand Rapids Fireball	Michigan, U.S.A.	8/27/68
54–68	Pocahontas Fireball	Iowa, U.S.A.	10/21/68
62–68	Ithaca Fireball	New York, U.S.A.	12/9/68
65–68	Arizona Fireball	Arizona, U.S.A.	12/14/68
3–69	Oita-Tokyo Fireball	Japan	1/7/69
5–69	Athens Fireball	Greece	1/16/69
7–69	Cedar Rapids Fireball	Iowa, U.S.A.	1/26/69
11–69	Pueblo de Allende Meteorite Shower	Mexico	2/8/69
14–69	Houston Fireball	Texas, U.S.A.	2/6/69
21–69	Malawi Fireball	Africa	1/21/69
34–69	Greenville Fireball	South Carolina, U.S.A.	3/31/69
37–69	Napa Fireball	California, U.S.A.	4/15/69
40–69	Belfast Meteorite Fall	Ireland	4/25/69
42–69	Houston Fireball II	Texas, U.S.A.	5/3/69
59–69	Iowa Fireball	Midwest U.S.A.	6/5/69
66–69	Araraquara Fireball	Brazil	6/11/69
72–69	Indiana Fireball	Midwest U.S.A.	6/26/69
80–69	Tunis Fireball	Tunisia	7/6/69
99–69	Ohio-Indiana Fireball	Ohio, Indiana, U.S.A.	8/5/69
102–69	Las Cruces Fireball	New Mexico, U.S.A.	8/9/69
104–69	North Island Fireball	New Zealand	6/3/69
120–69	Police Meteorite Fall	Czechoslovakia	9/16/69
124–69	Prince George Meteorite Fall	Canada	8/21/69

Event No.	Name	Location	Date
125–69	Murchison Meteorite Fall	Australia	9/28/69
126–69	Ozarks Fireball	Kansas, Missouri, U.S.A.	10/9/69
2–70	Lost City Meteorite Fall	Oklahoma, U.S.A.	1/8/70
8–70	Venezuelan Meteorite Fall	Venezuela	1/18/69
50–70	Holmes Fireball	Colorado, Wyoming, U.S.A.	5/18/70
56–70	Naini Tal Sudden Sky Brightening	India	6/9/70
61–70	Walaga Meteorite Fall	Ethiopia	6/19/70
65–70	North Haledon Fireball	New Jersey, U.S.A.	8/2/70
67–70	Ohio Fireball	Ohio, U.S.A.	8/8/70
69–70	Mozambique Channel Fireball	Mozambique	8/6/70
79–70	Lake Mills Fireball	Wisconsin, U.S.A.	9/10/70
81–70	Colorado-Nebraska Fireball	Colorado, U.S.A.	9/21/70
83–70	Stuttgart Fireball	Germany	8/27/70
91–70	Malakal Meteorite Fall	Sudan	8/10/70
94–70	Dwaleni Meteorite Fall	Swaziland	10/12/70
110–70	Mount Riffler Fireball	Austria-Germany	11/24/70
7–71	Kiffa Meteorite Fall	Mauritania	10/23/70
11–71	Wurzburg Fireball	Germany	1/17/71
12–71	Sakakah Fireball	Saudi Arabia	1/8/71
13–71	Princeton Fireball	Ohio, Indiana, Kentucky, U.S.A.	2/2/71
26–71	Ludwigshafen Fireball	Germany	2/23/71
32–71	Hawaii Fireball	Hawaii, U.S.A.	3/29/71
35–71	Wethersfield Meteorite Fall	Connecticut, U.S.A.	4/8/71
51–71	Gwarzo Meteorite Fall	Nigeria	4/25/71
58–71	Mawson Meteor Trail	Antarctica	5/9/71
64–71	Western Ohio Fireball	Ohio, U.S.A.	6/18/71
79–71	Haverö Meteorite Fall	Finland	8/2/71
24–72	Bramberg Fireball	Austria	3/21/72
25–72	Buedingen Fireball	West Germany	3/21/72
26–72	Marburg Fireball	West Germany	3/22/72
32–72	Lunar Meteorite Impact	Moon	5/13/72
47–72	Rankin Inlet Fireball	Canada	7/21/72
50–72	Alberta Fireball	Western U.S. and southwestern Canada	8/10/72
51–72	Arizona Fireball	Arizona, U.S.A.	8/8/72

Urgent Anthropological and Archaeological Events

Event No.	*Name*	*Location*	*Date*
51–68	Chinandega Footprints	Nicaragua	10/68
2–69	Santa Fe Prehistoric Camel Tracks	New Mexico, U.S.A.	1/69
6–69	Wama Stone Ax Tribe Discovery	Surinam	6/8/69
58–69	Mekong Valley Flooding	Southeast Asia	1971–86
61–69	Missouri Mastodon Fossil Find	Missouri, U.S.A.	6/4/69
75–69	Uthai Thani Site Discovery	Thailand	6/14/69
77–69	Sacsahuamán Miniature Sculpture Find	Peru	6/14/69
96–69	Colombian Tribe Discovery	Colombia	1969
106–69	Guyana Archaeological Find	Guyana	1969
30–70	Guajaki Tribe Discovery	Paranambu, Paraguay	3/30/70
67–71	Tasaday Manubo Discovery	Philippines	6/7/71

Bibliography

"Amphitheater," Encyclopaedia Britannica (1946 ed.), Vol. I, p. 847.

Annual Report 1971. Cambridge, Massachusetts: Smithsonian Institution, Center for Short-Lived Phenomena, Office of Environmental Sciences, May 1972.

Annual Report 1970. Cambridge, Massachusetts: Smithsonian Institution, Center for Short-Lived Phenomena, May 1971.

Annual Report 1969. Cambridge, Massachusetts: Smithsonian Institution, Center for Short-Lived Phenomena, 1 February, 1970.

Annual Progress Report 1968. Cambridge, Massachusetts: Smithsonian Institution, Center for Short-Lived Phenomena, 1 February, 1969.

"Appalachian Squirrel Migration—Appalachian Mountain Areas, September–October, 1968." Smithsonian Institution, Center for Short-Lived Phenomena Report. Event Report revised November 4, 1968.

"Bay of Bengal Delta, India," Encyclopaedia Britannica (1946 ed.), Vol. XV, p. 215.

"Bengalis Building an Anti-flood Earthworks," New York *Times,* May 20, 1970.

"Campania," Encyclopaedia Britannica (1946 ed.), Vol. IV, p. 674.

"Center Watches Nature's Quirks," New York *Times,* December 10, 1969.

Crocker, William H., and Schoen, Ivan L. "Notes on the Discovery of the Akurijo Stone Axe Tribe, Surinam." Smithsonian Institution, Center for Short-Lived Phenomena Report, February 20, 1969.

"Crowded East Pakistan a Low-Lying Delta Area," New York *Times,* November 17, 1970.

"Cyclone a 'Heat Engine' Over Vast Ocean Area," New York *Times,* November 16, 1970.

"East Pakistan Leader Voices Secession Threat," New York *Times,* November 27, 1970.

"East Pakistan Leaders Assail Yahya on Cyclone Relief," New York *Times,* November 24, 1970.

Elizalde, Manuel, Jr., and Fox, Dr. Robert B. "The Tasaday Forest People." Smithsonian Institution, Center for Short-Lived Phenomena Report (Data Paper), July 1971.

"Event Notification Cards, 1972." Cambridge, Massachusetts: Smithsonian Institution, Center for Short-Lived Phenomena. Nos. 1341, 1350, 1351, 1359, 1364, 1365, 1366, 1370, 1380, 1383, 1388, 1392, 1394, 1400.

Flyger, Vagn. "The 1968 Squirrel 'Migration' in the Eastern United States." Smithsonian Institution, Center for Short-Lived Phenomena Report, January 20, 1969.

"Going Back into Time," New York *Times,* April 9, 1972.
"Great Frog War, The," *Science Digest* (March 1971).
Guajaki Tribe, File on the: Letters, notes, transcriptions of conversations, Smithsonian Institution, Center for Short-Lived Phenomena, 1972.
"Indian Ocean," Encyclopaedia Britannica (1946 ed.), Vol. XII, p. 247.
Karen, Ruth. *The Seven Worlds of Peru.* New York: Funk & Wagnalls, 1969.
"Keeping up with Spiders, Bamboo, Sinking Islands," *International Herald Tribune,* January 14, 1970.
Kouprey, File on the: Letters, notes, documents, transcriptions of conversations, Smithsonian Institution, Center for Short-Lived Phenomena, 1972.
"Lucrinus Lacus," Encyclopaedia Britannica (1946 ed.), Vol. IV, p. 467.
"Malaysian Frog War and Other Phenomena, The," *Wall Street Journal,* June 28, 1971.
"Manila Gives Land to Cave Dwellers," New York *Times,* April 7, 1972.
"Mouse," Encyclopaedia Britannica (1972 ed.), Vol. XV, p. 963.
"Pakistan Death Toll 55,000; May Rise to 300,000," New York *Times,* November 17, 1970.
"Pakistan Relief Is Needed Badly," New York *Times,* November 22, 1970.
"Pakistan's Leader Hailed in Peking," New York *Times,* November 11, 1970.
"People Still Dying Because of Inadequate Relief Job," New York *Times,* November 20, 1970.
"Pozzuoli," Encyclopaedia Britannica (1972 ed.), Vol. XVIII, pp. 411ff.
———, Encyclopaedia Britannica (1946 ed.), Vol. XVIII, p. 398.
"Puteoli," Encyclopaedia Britannica (1946 ed.), Vol. XVIII, p. 783.
Rodman, Selden. *The Peru Traveler.* New York: Meredith Press, 1967.
"San Fernando, California, Earthquake of February 9, 1971, The," *Geological Survey Professional Paper 733.* U. S. Geological Survey and the National Oceanographic and Atmospheric Administration. Washington, D.C.: U. S. Government Printing Office, 1971.
Schoen, Ivan L. "Report of the Emergency Trip Made by the West Indies Mission to the Akurijo Indians, June 1971." Smithsonian Institution, Center for Short-Lived Phenomena Report, July 1971.
———. "Report on the Second Contact with the Akurijo Stone Axe Tribe, Surinam, September 1968." Smithsonian Institution, Center for Short-Lived Phenomena Report, February 4, 1969.
Sullivan, Walter. "East Pakistan Cyclone May Be the Worst Catastrophe of the Century," New York *Times,* November 22, 1970.
———. "Malaysian Frog War Really Love-in," New York *Times,* June 9, 1971.
———. "The Ups and Downs of Pozzuoli—a Volcanic Mystery," New York *Times,* March 29, 1970.
"Thousands of Pakistanis Killed by Tidal Waves," New York *Times,* November 15, 1970.
"Toad," Encyclopaedia Britannica (1972 ed.), Vol. XX, p. 42.
"Toll in Pakistan Put at 16,000," New York *Times,* November 16, 1970.
"U.S. to Use Tanker as Bengali Silo," New York *Times,* May 25, 1972.
"Volcano," Encyclopaedia Britannica (1972 ed.), Vol. XXIII, pp. 102ff.
"Volcanoes," Encyclopaedia Britannica (1946 ed.), Vol. XXV, pp. 238ff.

"Yahya Directing Disaster Relief," New York *Times,* November 26, 1970.

"Yahya Khan Visits Survivors," New York *Times,* November 26, 1970.

Yohner, Art, Jr. "Contact with a New Group of Akurijo Indians of Surinam." Smithsonian Institution, Center for Short-Lived Phenomena Report, November 5, 1970.